Geography Skills for NCEA Level Two

Justin Peat

NELSON
A Cengage Company

Australia • Brazil • Japan • Korea • Mexico • Singapore • Spain • United Kingdom • United States

Geography Skills for NCEA Level Two
2nd Edition
Justin Peat

Cover designer: McCarn Design
Text designer: Smartwork Creative Ltd
Production controller: Siew Han Ong

Any URLs contained in this publication were checked for currency during the production process. Note, however, that the publisher cannot vouch for the ongoing currency of URLs.

© 2016 Cengage Learning Australia Pty Limited

For product information and technology assistance,
in Australia call **1300 790 853**;
in New Zealand call **0800 449 725**

For permission to use material from this text or product, please email **aust.permissions@cengage.com**

National Library of New Zealand Cataloguing-in-Publication Data
A catalogue record for this book is available from the National Library of New Zealand.

978 0 17 038934 1

Cengage Learning Australia
Level 7, 80 Dorcas Street
South Melbourne, Victoria Australia 3205

Cengage Learning New Zealand
Unit 4B Rosedale Office Park
331 Rosedale Road, Albany, North Shore 0632, NZ

For learning solutions, visit **cengage.co.nz**

Printed in Malaysia by Papercraft.
10 11 25 24

Contents

Geographic
concepts
and skills

The aim of Achievement Standard Geography 2.4 is to assess your ability to *apply geography concepts and skills to demonstrate an understanding of a given environment*. In essence, the standard seeks to test your understanding of a geographical environment through the:

1 understanding of geography concepts; and

2 use of skills and geographic conventions in the presentation and/or interpretation of information.

Building upon Achievement Standard Geography 1.4, Geography 2.4 places equal emphasis on the understanding of geographic concepts as it does the ability to apply geographic skills. However, to achieve the standard beyond Achievement level also requires that the use of skills and geographic conventions is done so to a high level of accuracy and that one's understanding of geography concepts is detailed. Moreover, to achieve Excellence requires a comprehensive understanding of an environment be demonstrated through the thorough application of geography concepts, supported by the use of geographic terminology and insight (Figure 1).

	Criteria	Explanatory Notes
Achieved	Apply geography concepts and skills to demonstrate **understanding** of a given environment involves:	• Showing understanding of geography concepts.
		• Using skills and geographic conventions in the presentation and/or interpretation of information.
Merit	Apply geography concepts and skills **with precision** to demonstrate **in-depth** understanding of a given environment involves:	• Showing **detailed** understanding of geography concepts.
		• Using skills and geographic conventions to **a high level of accuracy** in the presentation and/or interpretation of information.
Excellence	Apply geography concepts and skills with precision to demonstrate **comprehensive** understanding of a given environment involves:	• Showing a **thorough** understanding of geography concepts using **geographic terminology** and showing **insight**.

Figure 1 Achievement criteria for Geography 2.4

ISBN: 9780170389341

With this in mind, this book is designed to help build upon the understanding of environments you acquired through the successful completion of Achievement Standard Geography 1.4 last year. Alternatively, if you are a new student of NCEA Geography, this book will help you to understand how to apply the geography concepts or the 'big ideas' of geography, and skills necessary to demonstrate an understanding of any given environment.

Understanding the geographic concepts or the 'big ideas' of geography

As a student of NCEA Geography, you will be required to gain an understanding of how certain concepts or 'big ideas' underpin the knowledge and skills necessary to interpret and represent information about the world. Although not comprehensive, the list of concepts forms the basis of *all* learning in NCEA Geography (Figure 2).

Several Maori concepts also have relevance to Geography, including kaitiakitanga (to care for), taonga (physical or cultural resource) and hekenga (migration).

Environments

Environments may be natural and/or cultural. They have particular characteristics and features, which can be the result of natural and/or cultural processes. The particular characteristics of an environment may be similar to and/or different from another.

Perspectives

Perspectives are ways of seeing the world that help explain differences in decisions about, responses to, and interactions with environments. Perspectives are bodies of thought, theories or world-views that shape people's values and have built up over time. They involve people's perceptions (how they view and interpret environments) and viewpoints (what they think) about geographic issues. Perceptions and viewpoints are influenced by people's values (deeply held beliefs about what is important or desirable).

Processes

A process is a sequence of actions, natural and/or cultural, that shape and change environments, places and societies. Some examples of geographic processes include erosion, migration, desertification and globalisation.

Patterns

Patterns may be spatial: the arrangement of features on the earth's surface; or temporal: how characteristics differ over time in recognisable ways.

Interaction

Interaction involves elements of an environment affecting each other and being linked together. Interaction incorporates movement, flows, connections, links and interrelationships. Landscapes are the visible outcome of interactions. Interaction can bring about environmental change.

ISBN: 9780170389341

Change

Change involves any alteration to the natural or cultural environment. Change can be spatial and/or temporal. Change is a normal process in both natural and cultural environments. It occurs at varying rates, at different times and in different places. Some changes are predictable, recurrent or cyclic, while others are unpredictable or erratic. Change can bring about further change.

Sustainability

Sustainability involves adopting ways of thinking and behaving that allow individuals, groups and societies to meet their needs and aspirations without preventing future generations from meeting theirs. Sustainable interaction with the environment may be achieved by preventing, limiting, minimising or correcting environmental damage to water, air and soil, as well as considering ecosystems and problems related to waste, noise, and visual pollution.

Figure 2 The prescribed geographic concepts or 'big ideas'

Applying geographic skills and conventions

This year, your ability to apply geographic concepts and skills in the presentation and interpretation of information will be tested in the specific external examination standard Geography 2.4 and also in the internal research standard Geography 2.5.

In the external examination, you will be provided with a booklet of resources, which you will use to show your understanding and application of geographic skills and concepts. The booklet will include a variety of resources such as maps, tables, diagrams, photographs and opinions. The resources in the booklet will be unfamiliar to you but will generally be about a particular geographic issue, which could be from New Zealand or from an overseas setting.

Basic geographic skills

Visual

- Interpretation of photographs, cartoons or diagrams including age-sex pyramids and models such as a wind rose
- Interpreting and completing a continuum to show value positions

Mapping

- Use of six-figure grid references and latitude and longitude
- Compass direction and bearings
- Distance, scale, area calculation
- Location of natural and cultural features
- Determination of height, cross-sections
- Use of a key, précis map construction
- Recognition of relationships, application of concepts
- Interpretation of other geographic maps like weather maps, cartograms, choropleth maps

Graphing

- Interpretation and construction of bar graphs (single and multiple), line graphs (single and multiple), pie and percentage bar graphs, scatter graphs, dot distribution, pictograms and climate graphs

Tables

- Recognition of pattern
- Simple calculation such as mean, mode and conversion to percentages

Figure 3 Basic geographic skills

ISBN: 9780170389341

Although few new skills are introduced at Level 2, the complexity of the material in the examination resource booklet will be greater than the more basic resources provided at Levels 1 (Figure 3). This is because the focus of Geography 2.4 is for you to develop your ability to interpret resources and present answers to a higher level of accuracy. For example, you could be asked to construct a specific type of graph within a given space (e.g. a triangular or scatter graph) and to do so with utmost accuracy (Figure 4).

	Basic skills at NCEA Level 1	Complex skills at NCEA Level 2
Latitude and longitude	• Degrees and minutes only.	• Degrees, minutes and seconds.
Direction	• To nearest inter-cardinal point (eight-point compass).	• To the nearest 16-point compass direction.
Scale	• Simple linear scale measurement on a map. • Recognition of different scales.	• Changed ratio scale with size. • Converting linear to ratio or vice versa. • Use of other scales apart from distance, i.e. time.
Graphing	• Scatter graph (interpretation or completion only).	• Triangular graphs. • Cumulative graphs. • Scatter graph construction. • Positive/negative. • Statistical mapping. • Proportional circle maps.
Tables	• All statistics given are used.	• Not all statistics given may be necessary for completion of tasks. • Percentage change calculations.

Figure 4 Progression of skills from NCEA Level 1 to Level 2

Examination ready

Because the complexity of examination questions increases at Level 2, one or two resources from the resource booklet may need to be used to answer a question. Similarly, instruction or command words used in Level 2 examination questions will also reflect the expectation of longer written explanations such as the need to *justify* or *evaluate* rather than just *describe* or *explain*, as was often required at Level 1. Recognising and understanding the requirements of the command word in a question is essential to answering it successfully (Figure 5).

Command word	Meaning	Example question
Analyse	Break down and interpret data or information in order to bring out the essential ideas.	Analyse the pattern of internal migration in a country.
Annotate	Add brief notes to a diagram or graph.	Draw an annotated diagram to show the key features of a coastal environment.

(Table continued over page.)

ISBN: 9780170389341

Command word	Meaning	Example question
Compare	Give an account of the similarities between two or more items or situations, referring to all of them throughout.	Compare coastal landforms on Auckland's west and east coasts.
Compare and contrast	Give an account of the similarities and differences between two or more items or situations, referring to all of them throughout.	Compare and contrast the tourist facilities in two different cities.
Construct	Display information in a diagrammatic or logical form.	Construct a climate graph to illustrate the data in the table.
Contrast	Give an account of the differences between two or more items or situations, referring to them throughout.	Contrast the effectiveness of anti-natalist and pro-natalist population policies.
Define	Give the precise meaning of a word, phrase, concept or physical quantity.	Define the Maori concept of kaitiakitanga.
Describe	Give a detailed picture of a given situation, event, pattern, trend, process or feature.	Describe the trend in population growth shown on the graph.
Discuss	Give a considered and balanced review that includes a range of arguments or factors. Opinions or conclusions should be presented clearly and supported by examples.	'Floods are more hazardous than earthquakes.' Discuss this statement.
Draw	Represent using a labelled, accurate diagram or graph using a pencil and a ruler. Diagrams should be drawn to scale, and graphs should have labelled axes, correctly plotted (if relevant) points joined in a line.	Draw a diagram to illustrate the features of a natural landscape.
Estimate	Give an approximate value.	Estimate the changes in the life expectancy rate over time.
Evaluate	Make an appraisal weighing up the strengths and weaknesses.	Evaluate the sustainability of tourism development in Queenstown.
Examine	Consider an argument or concept in a way that uncovers assumptions and interrelationships of an issue.	Examine why most countries want to reduce their dependence on oil.
Explain	Give a detailed account including reasons or causes.	Explain the relationship between fertility and female literacy.
Identify	Give an answer from a number of possibilities.	Identify the year in which there is an anomaly on the graph.
Justify	Give valid reasons or evidence/examples to support an answer or conclusion.	Justify the position taken by anti-globalisation movements.
Label	Add labels to a diagram.	Label the main features on the diagram.
Outline	Give a brief account or summary.	Outline two changes in literacy rates shown in the graphs.

ISBN: 9780170389341

Command word	Meaning	Example question
State	Give a specific name, value or other brief answer — no need for explanation or calculation.	State the three components that are used to calculate the Human Development Index.
Suggest	Propose a solution, hypothesis or other possible answer.	Suggest possible reasons for the changes in oil consumption between 1970 and 2010 shown on the graph.
To what extent	Consider the strengths/merits of an argument or concept. Opinions and conclusions should be presented clearly and supported by examples and sound arguments.	To what extent has climate influenced the environment of the Atacama Desert?

Figure 5 Command terms

Finally, as in the external examination, you are advised to use coloured pencils (black, blue, green, red, brown, and yellow), a calculator, and a ruler where appropriate when answering the questions in this booklet. You should use coloured pencils when constructing diagrams and maps. However, labels and annotations on these diagrams and maps must be in pen. Also, note that in the external examination, written work done in pencil will not be eligible for reconsideration.

ISBN: 9780170389341

Thematic mapping techniques

Knowing how to interpret a map correctly is a very important skill, as it provides us with information about places and helps us to identify patterns and changes in a landscape.

Thematic maps show the distribution (i.e. concentration, dispersal or flow) of geographic phenomena very effectively; they have particular geographic themes and are usually produced for specific audiences. There are many different types of thematic map, each differing according to its use and purpose. Generally, geographic information can be mapped in five ways:

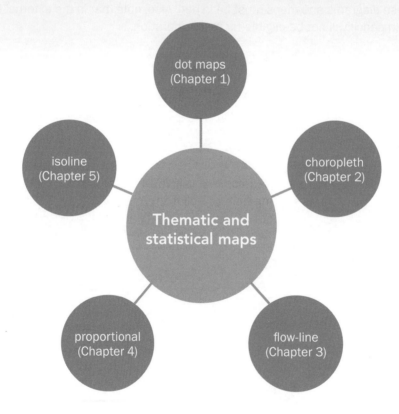

The amount of detail contained within a map will vary according to its purpose and scale. However, most maps have the same elements in common:

- a title stating the purpose of the map
- direction of orientation (i.e. north point)
- a scale (a ratio which compares a measurement on the map to the actual distance between locations identified on the map)
- a key or legend
- a border
- an indication of absolute location (e.g. a grid reference or latitude and longitude).

ISBN: 9780170389341

1 | Dot maps

As its name suggests, a dot map uses dots to show the distribution (or relative density) of geographic phenomena between different regions on a map. Each dot on a dot map represents a single feature or, in most cases, a quantity or number value. When viewed as a whole, a dot map will give its reader an impression of the overall distribution (or spread) of geographic phenomena throughout the area being mapped.

Dot maps have a wide variety of uses. For example, in medical geography dot maps have been used to show the spread of disease in a rural community, while in physical geography they have been used to show the pattern of earthquakes across a region. However, dot maps are most commonly used to compare population densities across regions.

The dot map in Resource 1.1 shows the degree of racial segregation (separation) in Los Angeles, USA. Using census data, the creator of this map assigned a value of 25 people to each dot (i.e. 1 dot = 25 people). After positioning each dot within the location of the census unit employed, the dots were then colour coded according to the dominant race at each location: White was colour coded pink; Black was coded blue; Hispanic coded orange; and Asian coded green.

Pattern of racial segregation in Los Angeles, USA

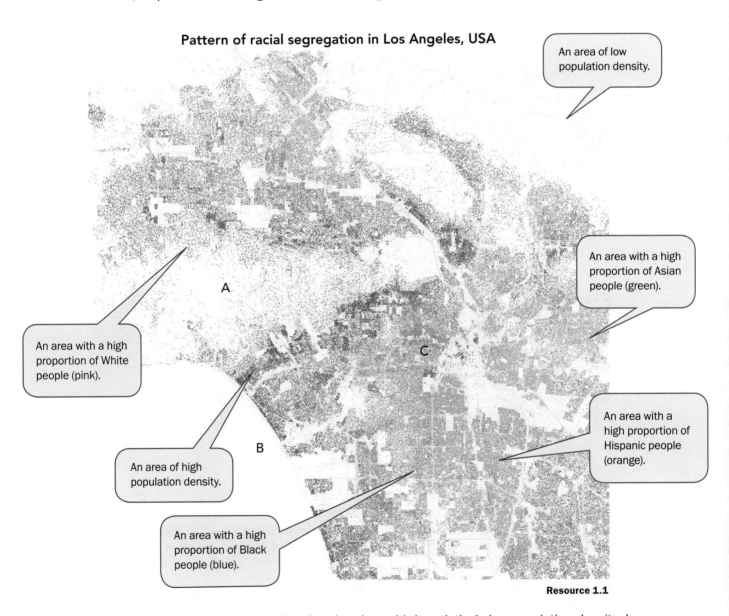

Resource 1.1

An analysis of Resource 1.1 shows that Los Angeles, with its relatively low population density, has some large blended neighbourhoods and several smaller, racially homogeneous neighbourhoods.

ISBN: 9780170389341

To interpret a simple dot map, follow the steps below:

1 Identify the geographic feature or phenomena being mapped.

2 Verify the dot value. This can be done by reading the map's legend.

3 Identify the scale of the administrative regions shown on the map (i.e. does the map show neighbourhoods, census areas, states or countries?).

4 Calculate the total value of features in each area of the map.

5 Describe the distribution of the feature both within and between different areas of the map. Resource 1.2 lists examples of some of the common patterns found on a map. The terms associated with each pattern can be used to describe the distribution of the mapped feature.

Common patterns found on a map

Cluster
A group of people or features positioned or occurring close together

Scattered or dispersed
Occurring or distributed over widely spaced and irregular intervals

Grid
A network of regularly spaced lines that cross one another at right angles

Linear
Arranged in or extending along a straight or nearly straight line
(e.g. a river or coastline)

Concentric
Circles that share the same centre, the larger often completely surrounding the smaller

Radial
Lines running directly from a centre point (e.g. town or city centre) to an outlying location

Resource 1.2

Although dot maps are usually used to illustrate the spatial distribution of one particular phenomenon (e.g. population density), multi-variable dot maps employ different colours to compare the distribution of related phenomena, or sub-groups within the same phenomena. Resource 1.3, for instance, shows how a multi-variable map can be used to compare changes in Auckland's resident population over time according to age.

Be aware, dot maps should not be mistaken for proportional symbol maps (Chapter 4) which sometimes look similar. That is because the symbols used in proportional symbol maps are always scaled in size according to the magnitude of data each symbol represents, whereas in dot maps they are not.

ISBN: 9780170389341

Change in census usually resident population, 2001 to 2013

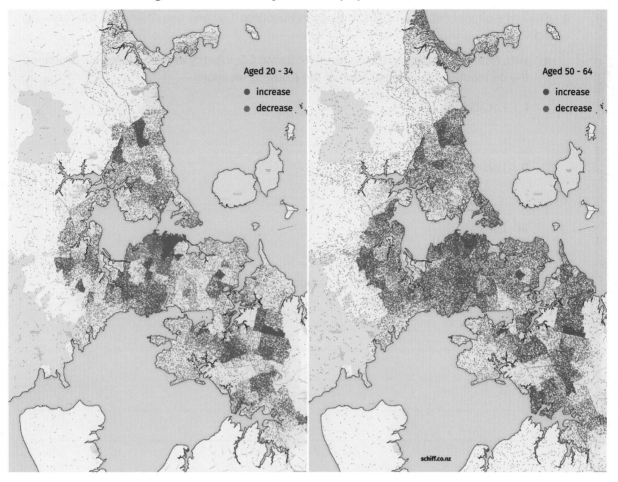

Resource 1.3 Change in Auckland inter-census resident population

To construct a dot map, follow the steps below:

1 Obtain a base map of the area or administrative regions you wish to show.

2 Study the data to be mapped and decide on a dot value. A rounded number should be chosen for the dot value to aid map interpretation.

3 Determine the number of dots required for each administrative region.

4 Decide on an appropriate dot shape and size. As a general guide, dots that are too small produce an overly sparse dot pattern which is unnecessarily precise, while dots that are too large produce excessively dense dot patterns, which can mask the distribution of the mapped phenomenon. As a single dot will represent a set value, for clarity of presentation it is important that the dot size remains consistent throughout the map.

5 Place the correct number of dots within each administrative boundary as determined in Step 2.

ISBN: 9780170389341

1 Study the simple dot map of Los Angeles in Resource 1.1, and then complete the following activities.

    a With reference to the patterns listed in Resource 1.2, identify which pattern best describes the distribution of people at each of the following locations.

        i Location A_____

        _____

        ii Location B_____

        _____

        iii Location C_____

        _____

    b Describe the degree of racial segregation across the city of Los Angeles.

2 Study the multi-variable dot map in Resource 1.3, and then with the aid of a New Zealand atlas or a detailed map of Auckland, complete the following activities.

    a With reference to areas within Auckland, describe the intercensal change in:

        i residents aged 20-34 _____

        ii residents aged 50-64 _____

    b What evidence is contained within Resource 1.3 to support the idea that Auckland's population is ageing?

3 The city of Curitiba, the largest city in southern Brazil, is world-famous for its creative and cost effective approach to dealing with the basic urban issues of overcrowding, poverty and pollution. One of Curitiba's planning successes was the establishment of waste recycling and collection sites known locally as 'Cambio Verde'. It is here that the citizens of Curitiba are encouraged to recycle through incentives which enable poor citizens to exchange their metal and glass waste for fresh produce (Resource 1.4).

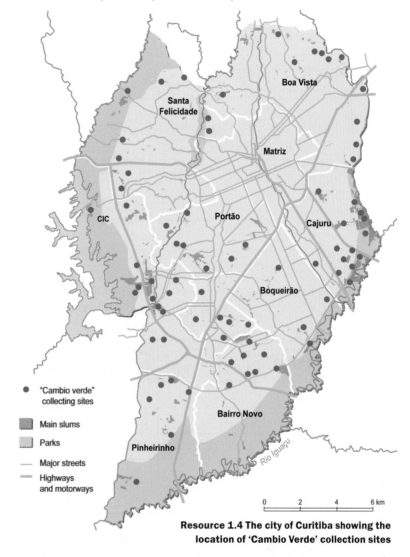

Resource 1.4 The city of Curitiba showing the location of 'Cambio Verde' collection sites

a Use the information in Resource 1.4 to complete the following activities.

   i Identify the area that has the highest concentration of Cambio Verde.

_____

   ii Identify the area that has the lowest concentration of Cambio Verde.

_____

   iii Identify the area that has the highest concentration of slums.

_____

   iv Identify the area that has the lowest concentration of slums.

_____

b Describe the relationship between the location of Cambio Verde and the location of slums. Suggest reasons for the relationship.

_____

_____

_____

_____

_____

_____

_____

_____

_____

_____

4 a Using the data below, construct a simple dot map of the regional population distribution of Maori in New Zealand on the map opposite. Use the ratio of 1 dot = 1000 people.

Region	Maori (2013 census)
Northland	55 200
Auckland	169 800
Waikato	96 100
Bay of Plenty	78 200
Gisborne	23 000
Hawke's Bay	39 500
Taranaki	20 400
Manawatu-Wanganui	49 200
Wellington	65 800
Tasman	3 870
Nelson	4 830
Marlborough	5 300
West Coast	3 600
Canterbury	47 900
Otago	16 300
Southland	12 750

Northland

Auckland

Waikato

Bay of Plenty

Gisborne

Taranaki

Hawke's Bay

Manuwatu-
Wanganui

Nelson

Wellington

Marlborough

Tasman

West Coast

Canterbury

Otago

Southland

b Describe the pattern shown in your map.

_____

_____

_____

_____

_____

_____

_____

_____

_____

_____

5 a Using the data in the table below, construct a simple dot map of under-five mortality in Africa on the map opposite.

Country	Under-five mortality rate per 1000 (2015)	Country	Under-five mortality rate per 1000 (2015)
Algeria	26	Madagascar	50
Angola	157	Malawi	64
Benin	100	Mali	115
Burkina Faso	89	Morocco	28
Cameroon	88	Namibia	45
Central African Rep.	130	Niger	96
Congo, Dem. Rep.	98	Nigeria	109
Congo, Rep.	45	Sierra Leone	120
Côte d'Ivoire	93	Somalia	137
Equatorial Guinea	94	South Africa	41
Gambia	69	South Sudan	93
Ghana	62	Sudan	70
Guinea	94	Swaziland	61
Lesotho	90	Tanzania	49
Liberia	70	Uganda	55
Libya	13	Zimbabwe	71

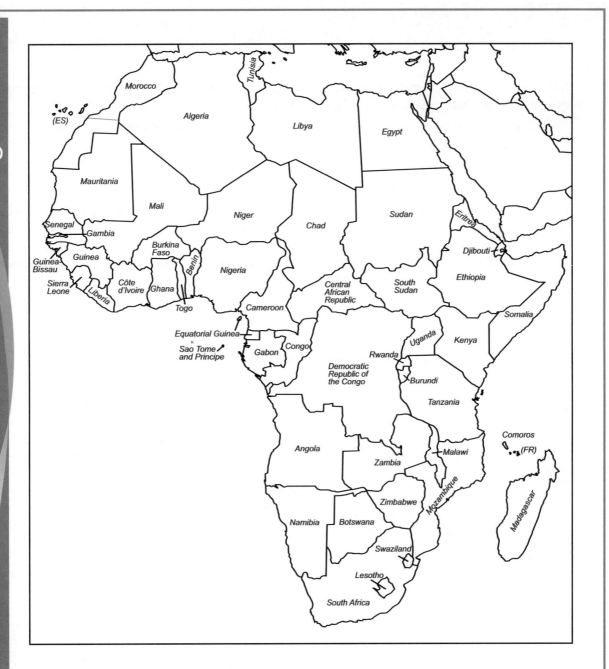

b Describe the pattern shown in your map.

_____

_____

_____

_____

_____

_____

_____

A choropleth map is a thematic map that uses proportional shading to reveal spatial patterns within geographic data. They are often used in geography to:

- represent values or quantities per unit area of land such as local authority boundaries or administrative regions
- compare relative densities of areas
- show change over time by comparing maps from different areas.

However, choropleth maps are most frequently used to show variations in population characteristics across a region or continent. The choropleth map in Resource 2.1, for example, illustrates differences in gender equality as measured by the number of girls per 100 boys who have access to primary education. The map shows that the greatest disparity between the number of boys and girls accessing primary education is in central Africa and West and Southern Asia.

Gender equality and empowerment of women: education status

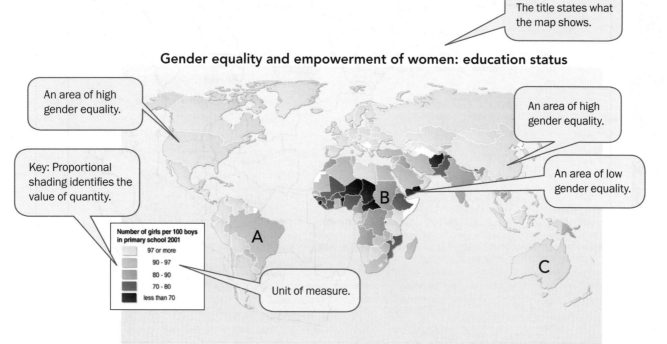

The title states what the map shows.

An area of high gender equality.

An area of high gender equality.

Key: Proportional shading identifies the value of quantity.

An area of low gender equality.

Number of girls per 100 boys in primary school 2001
- 97 or more
- 90 - 97
- 80 - 90
- 70 - 80
- less than 70

Unit of measure.

Resource 2.1

While choropleth maps show insufficient information for detailed analysis, they do allow for the quick observation of patterns and spatial variations and can act as a basis for further investigation.

To interpret a choropleth map, follow the steps below:

1 Identify the geographic feature or phenomena being mapped.
2 Verify the value of each shade used on the map. This can be done by reading the map's legend.
3 Identify the scale of the administrative regions shown on the map (i.e. does the map show suburbs, census areas, states or countries?).
4 Using the key as a guide, identify the areas of the map that share the same colour shading and therefore the same quantity or volume of the feature being mapped.
5 Describe the density or concentration of the feature both within and between different areas of the map.

While Resource 2.1 and Resource 2.2 (below) both use single colour progression to indicate more or less of a certain variable, it is not uncommon for choropleth maps to use a spectrum of contrasting colours or *bi-polar* colour progression when the range of data being mapped is extreme. With bi-polar colour progression, two contrasting colours are typically used to highlight both high and low values of a phenomenon (Resource 2.3).

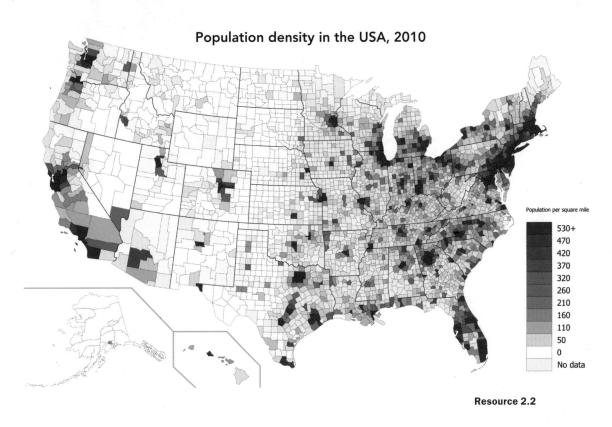

Population density in the USA, 2010

Population per square mile

530+
470
420
370
320
260
210
160
110
50
0
No data

Resource 2.2

To construct a choropleth map, follow the steps below:

1 Obtain a base map of the area or administrative regions you wish to show.

2 Examine the data to be presented and determine the range of data. This can be calculated by subtracting the lowest value in the data from the highest.

3 Divide the range by the number of categories you plan to use. Although most choropleth maps have four or five categories, more may be required if the data range is high. Regardless of the number of categories employed, ensure the value categories contain an even distribution of data (e.g. 1–100, 101–200, 201–300, 301–400, 401–500).

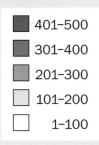

401–500
301–400
201–300
101–200
1–100

4 Assign a shade to each category. Typically, the darkest shade will be assigned to the highest value category while the lightest shade is assigned to the lowest value category.

5 Complete a category key on the map and sort the data into categories.

6 Shade in the administrative regions on your map according to your key.

ISBN: 9780170389341

1 Study Resource 2.3 and Resource 2.4, and then complete the following activities.

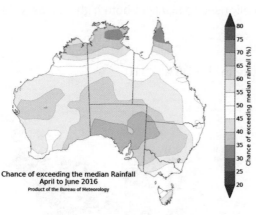

Chance of exceeding the median Rainfall
April to June 2016
Product of the Bureau of Meteorology

Chance of exceeding median rainfall (%)

Resource 2.3

Chance of exceeding the median Max Temp
April to June 2016
Product of the Bureau of Meteorology

Chance of exceeding median max temp (%)

Resource 2.4

a Identify the variable that is being mapped in:

i Resource 2.3

_____

ii Resource 2.4

_____

b What does the variation in shading show in:

i Resource 2.3

_____

ii Resource 2.4

_____

c With reference to Resource 2.3 and Resource 2.4, describe the expected pattern of weather for all of Australia for the period April to June, 2016.

_____

_____

_____

_____

_____

_____

_____

_____

_____

2 Study the choropleth map of gender equality in Resource 2.1, and then complete the following activities.

    a With reference to the patterns listed in Resource 2.1, identify the approximate number of girls per 100 boys enrolled in primary education at each of the following locations:

        i   Location A _____

        ii Location B _____

        iii Location C _____

    b With reference to different countries, describe the global pattern of gender equality in education.

_____

_____

_____

_____

_____

_____

_____

_____

_____

_____

_____

3 Study the choropleth map in Resource 2.2, and then complete the following activities.

    a Using an atlas to aid you, identify some of the regions of the USA that have a:

        i   High population density (more than 420 people per square mile).

_____

_____

        ii Low population density (50 to 160 people per square mile).

_____

_____

        iii Sparse population (less than 50 people per square mile).

_____

_____

b Write a paragraph describing the main differences in population density across the USA.

_____

_____

_____

_____

_____

_____

_____

_____

_____

_____

_____

4 a Using the data in the table below, construct a choropleth map to show the percentage of people living in urban areas in South-East Asia on the map opposite.

Country	Urban percentage of the population (2014)	Country	Urban percentage of the population (2014)
Cambodia	21	North Korea	61
China	54	Papua New Guinea	13
Indonesia	53	Philippines	44
Japan	93	South Korea	82
Laos	38	Thailand	49
Malaysia	74	Timor-Leste	32
Mongolia	71	Vietnam	33
Myanmar	34		

b Describe the pattern on the map.

_____

_____

_____

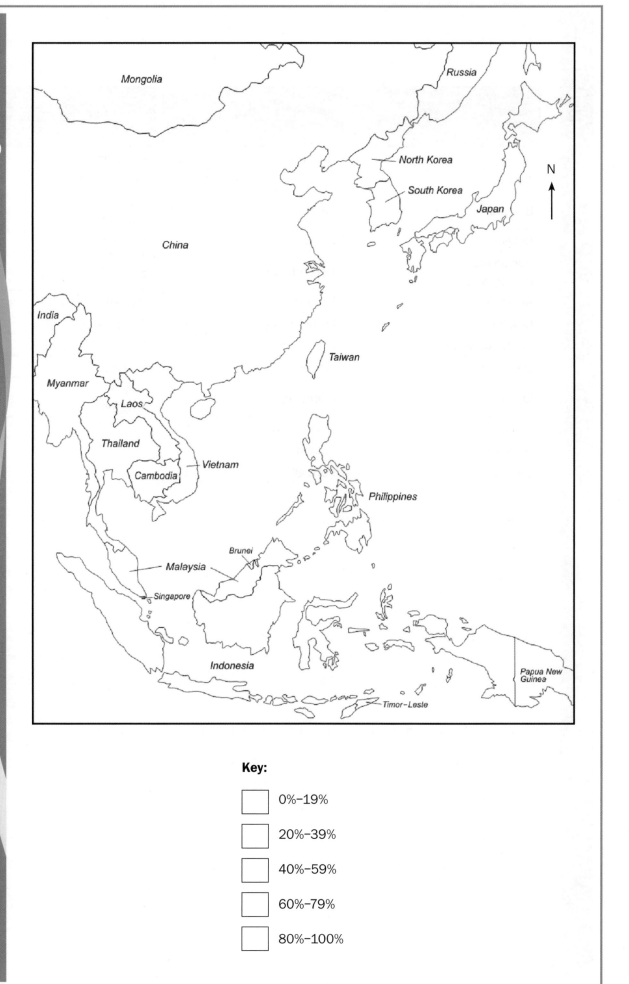

Key:

☐ 0%–19%

☐ 20%–39%

☐ 40%–59%

☐ 60%–79%

☐ 80%–100%

ISBN: 9780170389341

5 a Using the data in the table below, construct a choropleth map to show mobile phone connections per 100 people for selected countries in Europe on the map opposite.

Country	Mobile phone connections per 100 people (2015)	Country	Mobile phone connections per 100 people (2015)	Country	Mobile phone connections per 100 people (2015)
Albania	105	Hungary	118	Romania	106
Austria	152	Ireland	105	Serbia	122
Belarus	123	Italy	154	Slovak Republic	117
Belgium	114	Latvia	117	Slovenia	112
Bosnia and Herzegovina	91	Liechtenstein	109	Spain	108
Bulgaria	138	Lithuania	147	Sweden	128
Croatia	104	Luxembourg	149	Switzerland	137
Czech Republic	130	Macedonia	106	Turkey	95
Denmark	126	Monaco	88	Ukraine	144
Estonia	161	Montenegro	163	United Kingdom	124
France	101	Netherlands	116		
Germany	120	Poland	149		
Greece	110	Portugal	112		

b Describe the pattern on the map and suggest reasons for the pattern.

_____

_____

_____

_____

_____

_____

c What general conclusions can you draw from the map?

_____

_____

_____

_____

_____

_____

_____

Key:

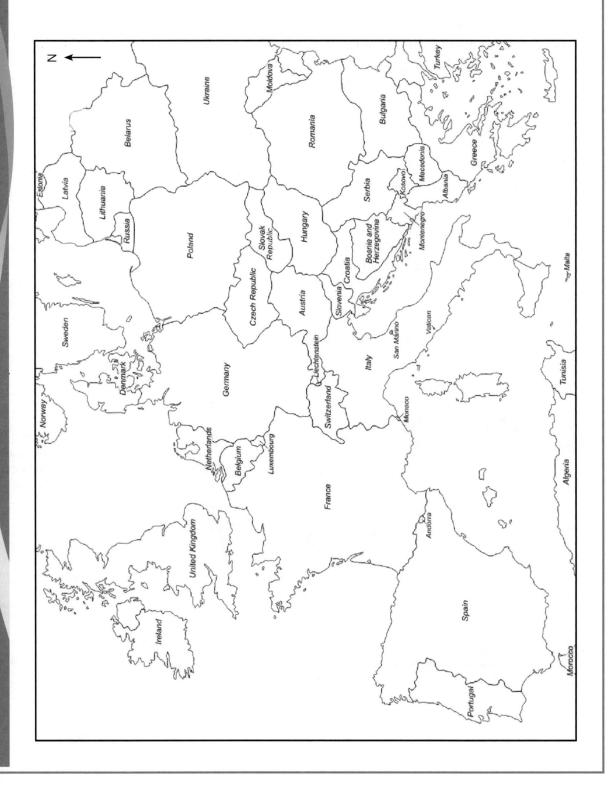

Flow-line maps are designed to show the flow or movement of geographic phenomena from one location to another such as the number of people in a migration, the volume of goods traded between regions, or water flows in a river basin.

When drawn well, flow-line maps allow the user to visualise the differences in magnitude or quantity of a range of flows. It achieves this by utilising arrows or lines of varying widths to represent the number of objects being transferred between the place of origin and the place of destination. To read a flow-line graph correctly, the user must understand how to interpret the map's flow-line scale, which determines the value of the map's flow lines.

To interpret a flow-line map, follow the steps below:

1 Identify the geographic feature or phenomena being mapped.

2 Verify the value of each line or arrow used on the map. This can be done by reading the map's legend.

3 Identify the scale of the administrative regions shown on the map (i.e. does the map show neighbourhoods, census areas, states or countries?).

4 Describe the direction and magnitude of the various movements of the geographic phenomena between different areas of the map.

The example of a simple flow-line map in Resource 3.1 shows the dollar value of timber products exported from countries of the Congo Basin (Angola, Cameroon, Central African Republic, Democratic Republic of the Congo, Equatorial Guinea, Gabon, Burundi and Rwanda) to China, North America, Europe and other parts of the world. The map shows that China is the largest importer of Congolese timber, followed by a cluster of European countries.

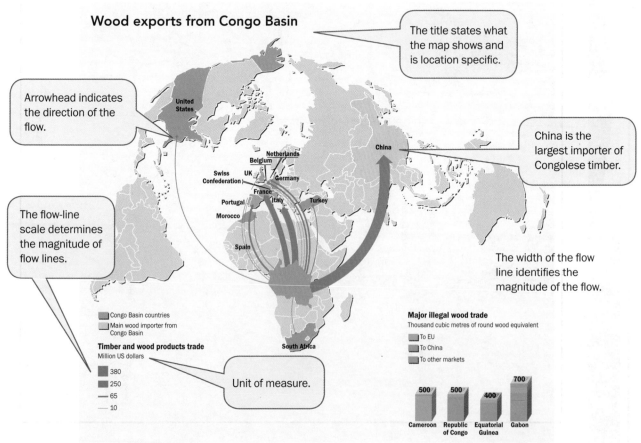

Wood exports from Congo Basin

The title states what the map shows and is location specific.

Arrowhead indicates the direction of the flow.

China is the largest importer of Congolese timber.

The flow-line scale determines the magnitude of flow lines.

The width of the flow line identifies the magnitude of the flow.

Congo Basin countries
Main wood importer from Congo Basin

Timber and wood products trade
Million US dollars

- 380
- 250
- 65
- 10

Unit of measure.

Major illegal wood trade
Thousand cubic metres of round wood equivalent

- To EU
- To China
- To other markets

	Cameroon	Republic of Congo	Equatorial Guinea	Gabon
	500	500	400	700

Resource 3.1

The flow-line map in Resource 3.2 summarises the magnitude of the Transatlantic Slave Trade which took place between 1500 and 1900. During a 400-year period, more than 15 million African men, women and children were forcibly transported to America and sold as slaves.

Overview of the Transatlantic Slave Trade, 1500–1900

Overview of the Slave Trade Out of Africa
Number of slaves
8,000,000
4,000,000
2,000,000
1,000,000
Width of routes indicates number of slaves transported

Resource 3.2 Overview of the Transatlantic Slave Trade, 1500-1900

To construct a simple flow-line map, follow the steps below:

1 Obtain a base map of the region you wish to show.

2 Study the data to determine how thick each flow line will need to be, to accurately communicate the magnitude of each flow. For example, 1 mm of line thickness could represent 100 data units.

3 Plot your data by drawing lines or arrows on the base map according to your predetermined scale. Position the tail of each flow line at the place of origin and add to it an arrowhead pointing to its destination.

4 Construct a key or legend to show the line scale, and the meaning of any symbols or shading used.

ISBN: 9780170389341

1 Study the flow-line map in Resource 3.3, and then complete the following activities.

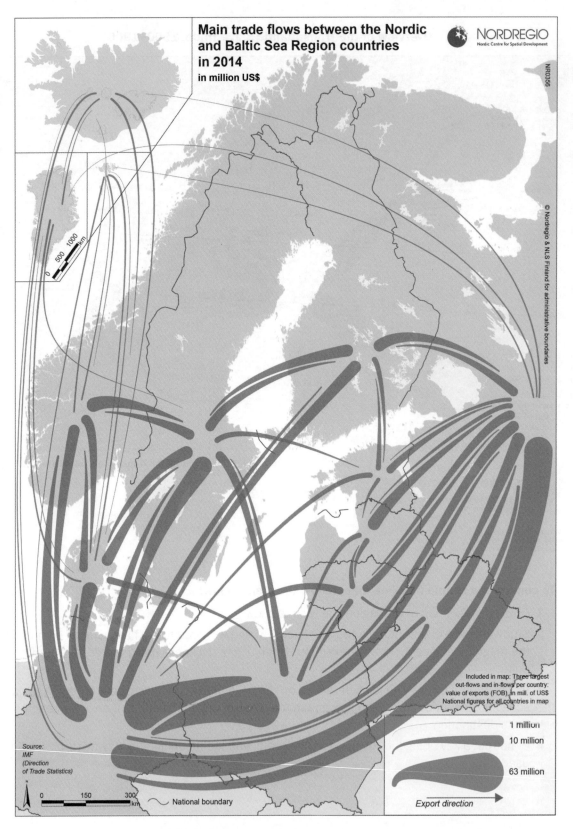

Main trade flows between the Nordic and Baltic Sea Region countries in 2014
in million US$

NORDREGIO
Nordic Centre for Spatial Development

Included in map: Three largest out-flows and in-flows per country: value of exports (FOB) in mill. of US$ National figures for all countries in map

Source:
IMF
(Direction of Trade Statistics)

1 million
10 million
63 million

National boundary

Export direction

Resource 3.3 Trade flows between countries in the Nordic and Baltic Sea region in 2014

a Using an atlas to guide you, identify and label onto Resource 3.3 the following countries:

i Denmark   **ii** Estonia   **iii** Germany

iv Finland   **v** Iceland   **vi** Norway

vii Poland   **viii** Russia   **ix** Sweden

b With reference to named countries and evidence from the map, outline the pattern of trade for:

i Estonia

_____

_____

_____

_____

_____

_____

ii Norway

_____

_____

_____

_____

_____

_____

iii Russia

_____

_____

_____

_____

_____

_____

_____

ISBN: 9780170389341

2 a Construct a flow-line map to show the movement of people between regions of New Zealand based on the censuses of 2008 and 2013.

Selected migration flows by regional council area (2008–2013 censuses)

Source region	Destination region	Net flow	Source region	Destination region	Net flow
Auckland	Bay of Plenty	7 770	Manawatu	Waikato	3 189
Auckland	Canterbury	6 264	Northland	Auckland	6 897
Auckland	Northland	8 736	Southland	Canterbury	2 412
Auckland	Otago	4 566	Southland	Otago	3 558
Auckland	Waikato	15 678	Waikato	Auckland	11 301
Canterbury	Auckland	8 796	Waikato	Bay of Plenty	7 821
Canterbury	Otago	7 467	Wellington	Auckland	9 201
Canterbury	Wellington	7 467			

b Describe in detail the pattern on your map.

_____

_____

_____

_____

_____

_____

_____

_____

c To what extent can it be argued that the population of New Zealand moving is north or south?

_____

_____

_____

_____

_____

_____

3 Study the simple flow-line map in Resource 3.1, and then complete the following activities.

a Estimate the value of timber exported from the Congo Basin to:

i China _____

ii France _____

iii USA _____

iv South Africa _____

v Portugal _____

b Describe the trade of timber from the Congo Basin.

_____

_____

_____

_____

_____

4 Study the flow-line map in Resource 3.2, and then complete the following activities.

    a Estimate the magnitude of the flow of slaves between Africa and:

        i North America _____

        ii South America _____

        iii Europe _____

        iv Asia _____

    b Outline the movement of slaves during the Transatlantic Slave Trade 1500-1900.

    c To what extent could the Transatlantic Slave Trade have been considered a global issue?

5 **a** Construct a flow-line map to show the flow of refugees resulting from the Syrian Civil War (2015–).

Country of asylum	Total refugees	Country of asylum	Total refugees	Country of asylum	Total refugees
Bulgaria	17 089	Jordan	1 265 000	Russia	5 000
Cyprus	3 185	Lebanon	1 500 000	Saudi Arabia	420 000
Egypt	118 512	Libya	26 672	Serbia (incl. Kosovo)	313 035
Greece	480 000	Macedonia	400 000	Turkey	2 715 789
Italy	2 451				

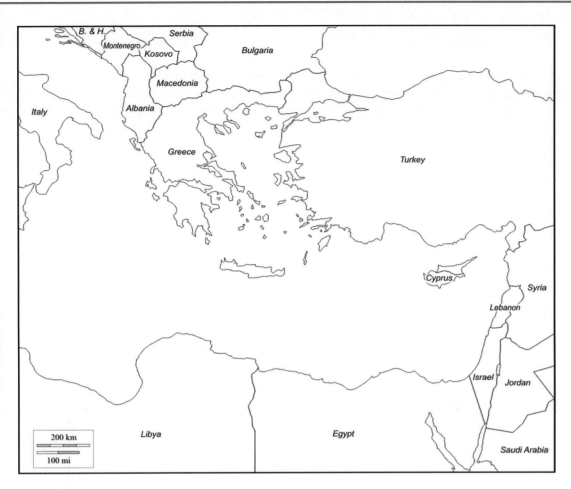

b Describe in detail the pattern shown in your map.

_____

_____

_____

_____

_____

_____

_____

_____

A proportional symbol map is a widely used thematic map which utilises symbols of different sizes to represent data associated with different locations on a map. Proportional symbol maps are frequently used in geography because they are easy to read, allowing for the simple but effective identification of spatial patterns.

As with all thematic maps, employing proportional symbol maps has its advantages and disadvantages.

Advantages:

- They are easy to interpret.
- Different symbols can be used to illustrate the distribution of multiple phenomena on the same map.
- They have the ability to show the different attributes of individual phenomena (Resource 4.4).

Disadvantage:

- Their interpretation is vulnerable to inaccurate perception of symbol size.

Almost any shape can be used in the construction of proportional symbol maps including circles, squares, triangles and bars. Regardless of the shape employed, the size and area of the symbol must be in proportion to the data value it is representing (Resource 4.1).

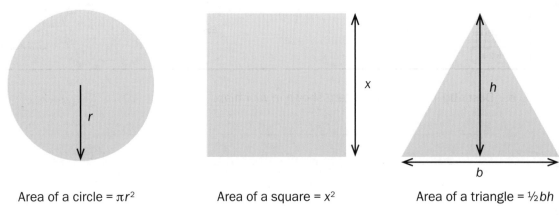

Area of a circle = πr^2 Area of a square = x^2 Area of a triangle = $\frac{1}{2}bh$

Resource 4.1 Calculating the area of commonly used proportional symbols

To interpret a proportional symbol map, follow the steps below:

1 Identify the geographic feature or phenomena being mapped.

2 Verify the value of each proportional symbol used on the map. This can be done by reading the map's legend.

3 Identify the scale of the administrative regions shown on the map (i.e. does the map show suburbs, census areas, states or countries?).

4 Calculate the total value of phenomena in each area of the map.

5 Describe the distribution of the phenomena both within and between different areas of the map.

In the example on the next page (Resource 4.2), proportional circles of different colours are used to show the number of people killed by flood events (red circles) versus the number of people generally affected by flood events (blue circles) in the Hindu Kush-Himalaya region from 2008 to 2010. Interpretation of the map reveals that most flood events occurred near major rivers, and that more people were affected by the floods than were killed.

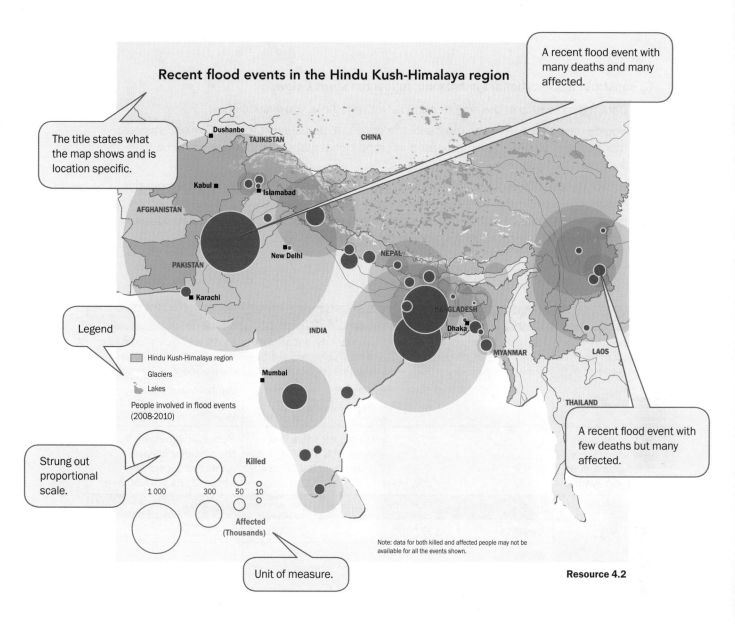

Resource 4.2

The unit of scale used in a proportional map can be displayed in the legend in three different ways: bar, nested or strung out (Resource 4.3).

Scaling of proportional circles

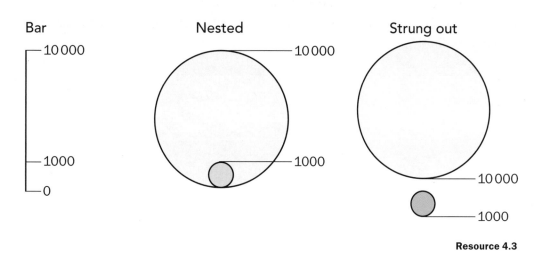

Resource 4.3

To construct a proportional symbol map, follow the steps below:

1 Obtain a base map of the area or administrative regions you wish to show.

2 Decide on a shape to use to represent the data you are planning to plot. Circles, squares, triangles and bars are commonly used in proportional maps as it is not too difficult to change their area to represent different values.

3 To plot the data correctly you will need to scale each symbol. Proportional circles are the most commonly used symbol as they are easy to draw and can be scaled to represent values by changing the radius. For instance, the circle symbol representing a city of 100 000 would be 10 times larger than the symbol for representing a town of 10 000.

Step 1	Step 2	Step 3	Step 4
Decide on the diameter of the largest symbol.	List of the elements and data values to be represented.	Calculate the square root of each value and use the results from this list in Step 4. $result = \sqrt{value}$	Use the following formula to calculate the size of each data symbol: symbol size = $\dfrac{result}{(largest\ result)} \times$ maximum symbol size
For this example, we chose 50 mm as the diameter of our largest symbol.		The largest value in our example is 1 377 200; its square root is 1173.54.	Hence, in our example, the diameter of each symbol would equal (result/1173.54) x 50 mm.

City	Population	√ population	Symbol size (mm)
Auckland	1 377 200	1173.54	50
Wellington	393 400		
Christchurch	380 900		
Hamilton	206 400		
Napier-Hastings	124 800		

4 Plot the data by drawing proportional symbols on the base map according to your predetermined scale. Take care to ensure the symbols do not overlap.

5 Construct a legend to show the proportional scale and meaning of any symbols or shading used (Resource 4.3).

ISBN: 9780170389341

1 Study the proportional symbol map in Resource 4.4, and then complete the following activities.

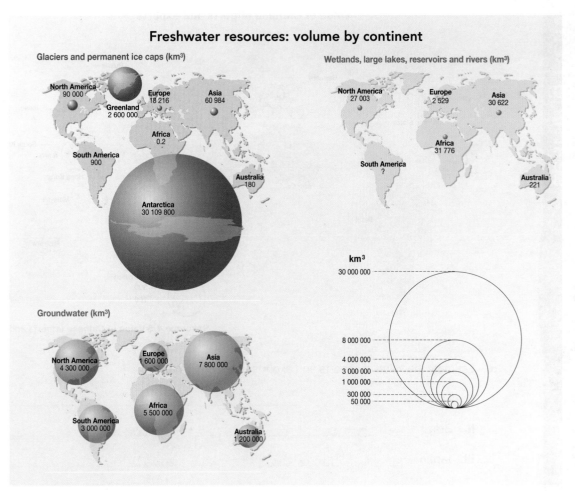

Resource 4.4

a State the volume of freshwater resources stored as ice in:

i North America _____

ii Africa _____

iii Antarctica _____

b State the volume of freshwater resources stored as groundwater in:

i Australia _____

ii Africa _____

iii Europe _____

c Identify which continent has the greatest volume of freshwater and which continent has the lowest volume.

_____

_____

2 Study the proportional symbol map in Resource 4.5, and then complete the following activities.

Value of Chinese imports and exports

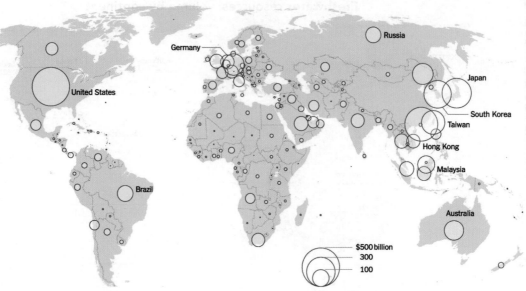

Resource 4.5 Value of Chinese imports and exports

a Estimate the value of imports and exports between China and:

i Australia _____

ii Brazil _____

iii Japan _____

iv New Zealand _____

v United States _____

b Suggest why China's trading (imports and exports) patterns may be regarded as global in scale.

_____

_____

_____

_____

_____

3 a With the aid of an atlas, construct a proportional symbol map to show the location and size of New Zealand's largest cities.

New Zealand city	Population (June 2015)
Auckland	1 454 300
Wellington	398 300
Christchurch	381 800
Hamilton	224 000
Napier – Hastings	130 800
Tauranga	129 700
Dunedin	117 400

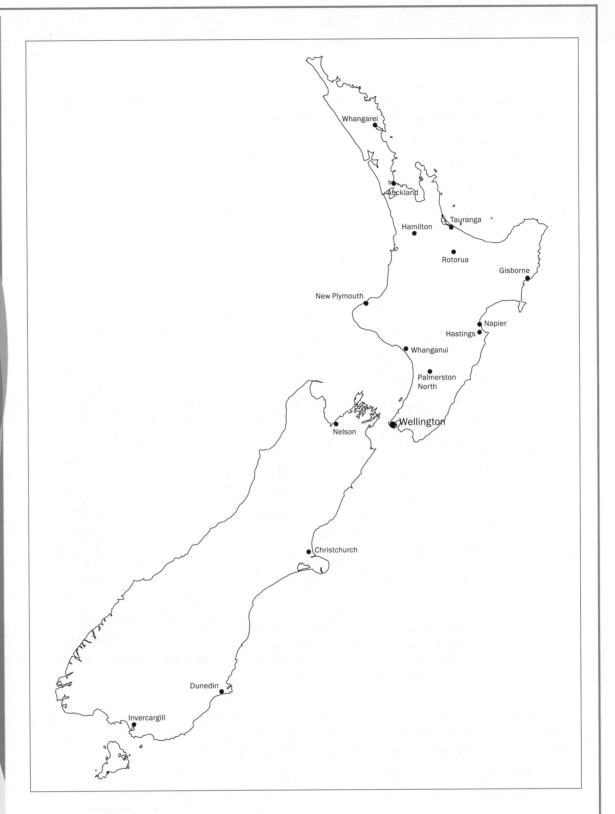

b Describe the pattern shown in your map.

_____

_____

_____

_____

Isoline maps are made up of lines that join points of equal value. Perhaps the most common of all isoline maps is the *synoptic* chart, or as it is more commonly known, the weather map. Weather maps use an array of symbols and lines called isobars to join together areas of equal air pressure.

As air pressure influences the type of weather experienced at a particular location, a general understanding of air pressure and how it relates to regional weather patterns is an invaluable skill to have.

Air pressure is measured in hectopascals (hPa) or millibars (mb). One hectopascal is equivalent to one millibar. In general, areas of high pressure (>1013 mb) are normally associated with clear skies and little wind. The letter H is normally used to identify an area of high pressure on a weather map. By contrast, areas of low air pressure (<1013 mb) are normally associated with rain, cloudy skies and strong winds. The letter L is most frequently used to identify an area of low pressure on a weather map.

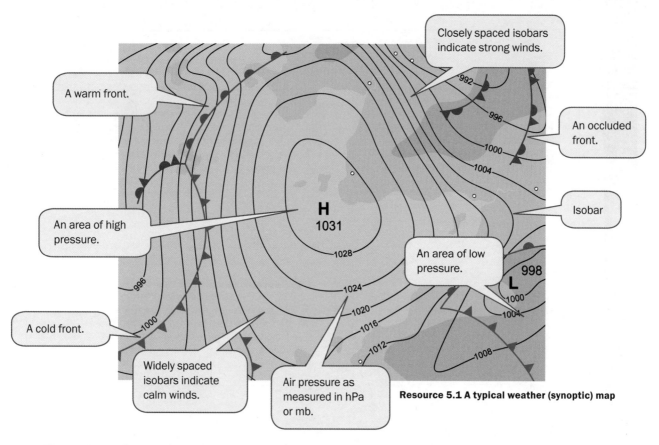

Resource 5.1 A typical weather (synoptic) map

The makers of weather maps employ symbolism to communicate the location of important weather features or phenomena to the reader. One important symbol found on a weather map is the *front*. Weather maps can include any one or more of four types of front:

- **Cold front:** The leading edge of a fast-moving mass of cooler air.
- **Warm front:** The leading edge of a slow-moving mass of warmer air.
- **Stationary front:** The boundary between two different air masses, neither of which is strong enough to overcome the other.
- **Occluded front:** When a fast-moving cold front overcomes a slow-moving warm front, an occluded front can form.

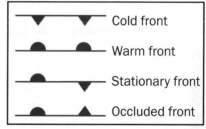

Resource 5.2

To interpret a weather (isoline) map, follow the steps below:

1 Identify pressure systems: Establish whether the pressure systems on the map represent areas of high (H) or low (L) pressure. Areas of high pressure (anticyclones) have enclosed isobars which increase in pressure towards their centre, while low pressure systems (depressions) contain isobars which decrease in pressure towards the centre.

2 Identify wind conditions: First establish which hemisphere (Northern or Southern) the area in the map is located. In the Southern Hemisphere wind flows clockwise around low pressure systems and anti-clockwise around high pressure systems. In the Northern Hemisphere the flow is the other way around. Isobars help identify the wind's direction but are not fool-proof. Winds generally flow towards the centre of low pressure systems and away from high pressure systems. Winds are strongest where isobars are closer together.

3 Identify frontal systems: As different types of front are associated with different weather patterns, the location and predicted path of a front will help establish the type of weather experienced at a particular location (Resource 5.2).

Learning Activities

1 Study the satellite photograph below and then complete the following activities.

  a Identify the type of weather system in the photograph. _____

  b Which hemisphere is the weather system located in? _____

  c Describe the flow of wind in the pressure system.

  _____

  _____

  _____

2 Study the following synoptic charts captured over four consecutive days, and then complete the following activities.

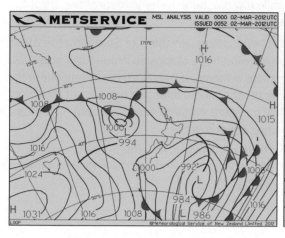

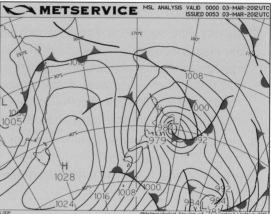

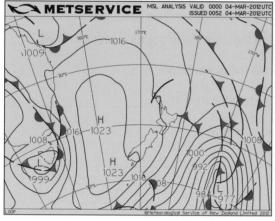

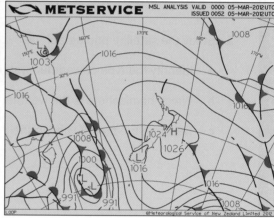

a Looking at the synoptic chart dated 2 March 2012, identify:

   i The pressure system to the north and northeast of New Zealand _____

   ii The pressure system to the west of the North Island _____

   iii The front to the west of the North Island _____

   iv The front to the south of the South Island _____

b Looking at the synoptic chart dated 3 March 2012, identify:

   i The lowest reading on the map _____

   ii The lowest air pressure reading over the North Island _____

   iii The highest air pressure recording over the Tasman Sea _____

   iv The air pressure range for the entire chart _____

c Looking at the synoptic chart dated 4 March 2012, describe the weather situation for New Zealand in terms of:

i Wind direction over the North Island _____

ii Wind direction over the South Island _____

3 Imagine you are a weather presenter on national radio whose task is to broadcast the weather for all of New Zealand for the four days following 1 March 2012. Write a narrative of what you would say if you were presenting the forecast on 1 March 2012. In your narrative describe the type of weather expected (wind strength and direction, expected rainfall) and the pressure systems involved.

_____

_____

_____

_____

_____

_____

_____

_____

_____

_____

_____

_____

_____

_____

4 Some media reports at the time described the weather events of the 2–4 March as a 'weather bomb'. With reference to the four synoptic charts opposite, explain why the media used this terminology.

_____

_____

_____

_____

_____

_____

ISBN: 9780170389341

Working with topographic maps

The most common map used by geographers is the *topographic* map. Topographic maps are detailed, accurate graphic presentations of features on the Earth's surface. Examples of features shown on topographic maps are given below.

Features shown on a topographic map

Cultural features
- roads
- buildings and urban areas
- transport networks
- place names

Vegetation features
- native and plantation forest
- vineyards and orchards
- scrub

Relief features
- mountains
- ridges and valleys
- slopes and depressions

Drainage features
- rivers
- lakes
- streams
- swamps

The main topographic map series for New Zealand is the Topo50 and the Topo250 series, produced and published by Land Information New Zealand (LINZ). Topo50 maps are produced to a scale of 1:50 000 and Topo250 maps are produced to a scale of 1:250 000. Both map series show geographic features in detail making them ideal for studies of the environment (opposite).

In Level 1 NCEA Geography you were introduced to many of the basic mapping skills. These skills included:

- direction and relative location (to the nearest of the eight inter-cardinal compass points)
- latitude and longitude (degrees and minutes only)
- six-figure grid references
- recognising scale
- cross-sections
- map symbols
- précis maps.

While these skills are still important in Level 2 NCEA Geography, some of them will become more complex. The more complex skills introduced at Level 2 include:

- direction and relative location (to the nearest of 16 compass points)
- 14-figure grid references
- latitude and longitude (degrees, minutes and seconds)
- scale (converting linear to ratio or vice versa)
- use of scales other than distance (i.e. time).

ISBN: 9780170389341

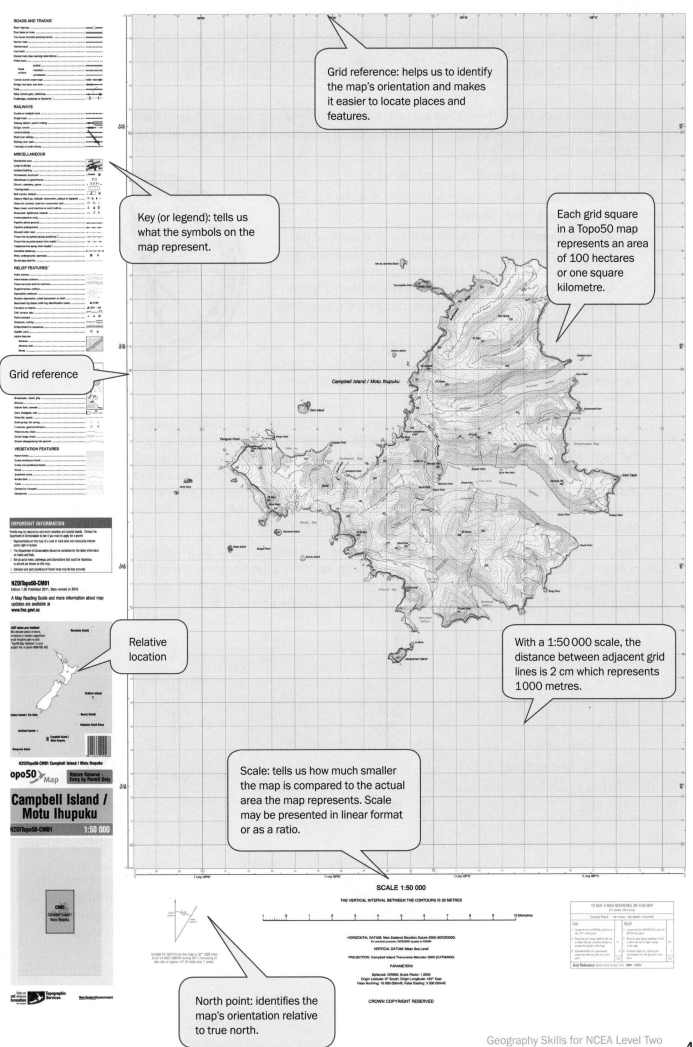

Grid reference: helps us to identify the map's orientation and makes it easier to locate places and features.

Key (or legend): tells us what the symbols on the map represent.

Each grid square in a Topo50 map represents an area of 100 hectares or one square kilometre.

Grid reference

Relative location

With a 1:50 000 scale, the distance between adjacent grid lines is 2 cm which represents 1000 metres.

Scale: tells us how much smaller the map is compared to the actual area the map represents. Scale may be presented in linear format or as a ratio.

North point: identifies the map's orientation relative to true north.

The notion of location (or the particular place where something is) is the most basic of all geographical concepts. It is therefore important that as a student of geography you refine the skill of locating places before mastering any other.

Location is usually expressed in one of two ways:

- **Relative location** describes the location of a place or feature as it relates to other features.

- **Absolute location** refers to the location of a point on the Earth's surface, referenced to a grid system, as it appears on a map. Of the many grid-referencing systems in use, the three that are important for students of Level 2 NCEA Geography are the six- and 14-figure grid reference systems, and latitude and longitude.

Relative location is usually expressed in one of two ways. The most common method is to describe *direction* according to the points of a compass, however, it can also be expressed as a *bearing* when a more precise reading is required.

In Level 1 NCEA Geography you would have been introduced to the four *cardinal* (main) points of the compass: north (N), south (S), east (E) and west (W); as well as the four *intermediate* points that form the eight-point compass rose. In Level 2 Geography you will need to also learn the additional eight points that form the 16-point compass rose. Sixteen-point compass roses are constructed by bisecting the points of the eight-point compass rose to come up with additional compass points, known as half-winds. The names of the half-winds are simply combinations of the cardinal and intermediate winds on either side e.g. north-northeast (NNE), east-northeast (ENE) etc.

As a compass is a circle made up of 360 degrees, its directions can also be expressed as an angle or bearing. Bearings can be measured on a map with the aid of a protractor and are always stated as three-figures measured clockwise from north (000°). For example, east is 090° clockwise of north, southeast is 135° and northwest is 315° clockwise of north (Resource 6.1).

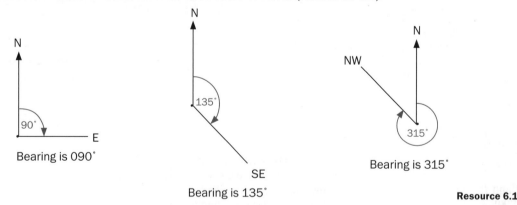

N
90°
E
Bearing is 090°

N
135°
SE
Bearing is 135°

N
NW
315°
Bearing is 315°

Resource 6.1

Of course, to measure bearing on a map, you first need to establish the direction of north. Most topographic maps will show grid north in the format of parallel north-south grid lines. Grid north differs from magnetic and true north in that it provides a fixed orientation to measure from. Most maps will show grid north (000°) pointing to the top of the page, however, exceptions to this rule exist so always confirm the map's orientation before undertaking any measures.

To use bearings to show direction, follow the steps below:

1 Using a ruler and pencil, rule a line joining the points *x* and *y*.

2 Place a clear protractor over the line *x-y* so that the 0° reading on the protractor points to grid north and the *x* point is positioned at the centre of the protractor.

3 Reading clockwise from 0°, state the angle that the *x-y* line intersects the measuring edge of the protractor. The angle measured is the bearing from point *x* to point *y*.

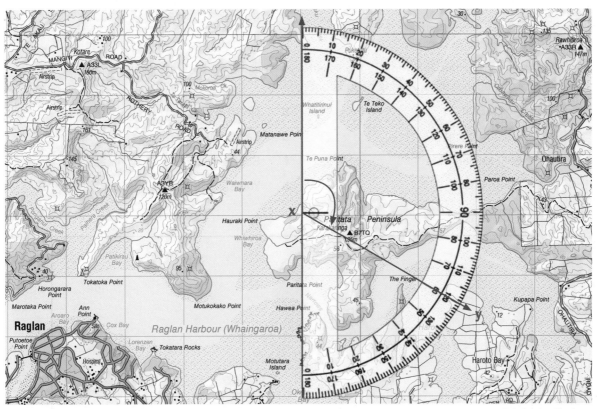

Resource 6.2 Raglan Harbour

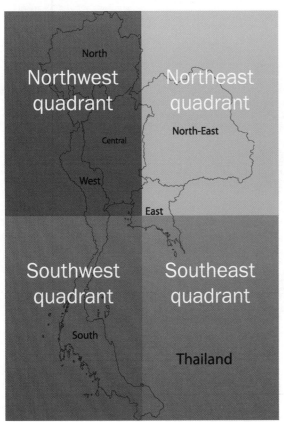

It is useful to be able to describe the relative location of features on a map by expressing their location in terms of compass quadrants. This is done by dividing the map into quarters and then naming each quarter according to the points of a compass (Resource 6.3).

Resource 6.3

ISBN: 9780170389341

1 Describe the difference between relative location and absolute location.

_____

_____

_____

_____

2 Compass direction is often used to identify the regions of a country. With the aid of an atlas, locate and label the following regions of England.

a	North East	**b**	North West
c	West Midland	**d**	East Midlands
e	East England	**f**	South West
g	South East	**h**	Yorkshire and the Humber
i	London		

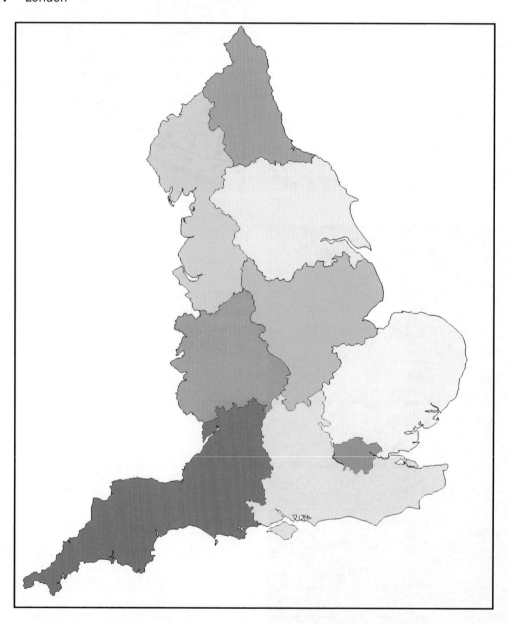

3 With reference to the map of the Mediterranean region below and the aid of a protractor, plot the route and then state the bearing you would travel if you were to fly from:

a Madrid to Barcelona _____

b Barcelona to Rome _____

c Rome to Tunis _____

d Tunis to Algiers _____

e Algiers to Madrid _____

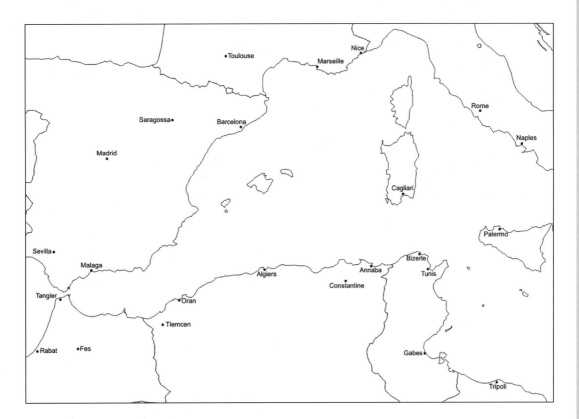

4 Study Resource 6.2 and then complete the following activities.

a State the straight line compass direction you would travel if you were to kayak from:

i Ann Point to Tokatara Rocks _____

ii Tokatara Rocks to Hawea Point _____

iii Hawea Point to Harongarara Point _____

iv Harongarara Point to Ann Point _____

b State the bearing direction you would travel if you were to kayak from:

i Ann Point to Harongarara Point _____

ii Harongarara Point to Hawea Point _____

iii Hawea Point to Tokatara Rocks _____

iv Tokatara Rocks to Ann Point _____

Maps are particularly useful to geographers when it comes to finding absolute location. The exact location of a feature on a topographic map, for example, is determined with grid lines. Grid lines are equally spaced vertical and horizontal lines drawn on a map and are called:

- **Eastings:** vertical grid lines that divide the map into columns from west to east (*x*-coordinate)
- **Northings:** horizontal grid lines that divide the map into rows from north to south (*y*-coordinate).

Eastings and northings intersect each other on a map to form grid squares. The grid squares are then used to help calculate unique six-figure grid references (GR) for individual features (Resource 7.1).

Six-figure grid references contain six digits. The first three digits of a six-figure grid reference refer to the easting, and the second three digits refer to the northing. The third and sixth digit is determined by dividing the easting and northing into tenths (Resource 7.2).

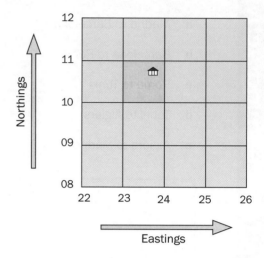

Resource 7.1

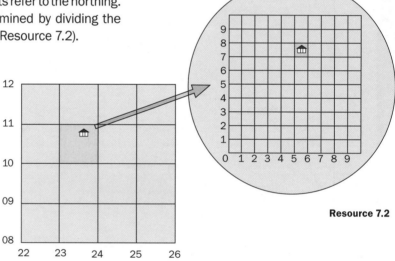

Resource 7.2

To give the absolute location of a feature using six-figure grid references for the example above (Resource 7.2), follow the steps below:

1. To determine the six-figure grid reference for the 🏛 symbol in Resource 7.2, identify the first vertical grid line to the left of the symbol. In the example above, the first grid line to the left is labelled *23* in the bottom margin. This is the easting.

2. Estimate the number of tenths eastward the symbol is from the easting on the left. In the example above, the symbol is located a distance of $\frac{6}{10}$ from the easting on the left. This gives a final easting reading of *236*.

3. Now identify the first horizontal grid line below the symbol. In the example above, the first grid line below is labelled *10* in the left margin. This is the northing.

4. Estimate the number of tenths northward the symbol is from the northing beneath. In the example above, the symbol is located a distance of $\frac{7}{10}$ from the northing beneath. This gives a final northing reading of *107*.

5. You have now calculated the six-figure grid reference for the 🏛 symbol. It should be written as GR 236107.

1   Study the map below and then complete the following task.

   a   State the six-figure grid reference for the following features:

      i   Lake Johnson _____

      ii   Sugar Loaf summit _____

      iii Skyline Chalet _____

      iv   Queenstown Hill _____

      v   Big Beach _____

      vi   Arthurs Point _____

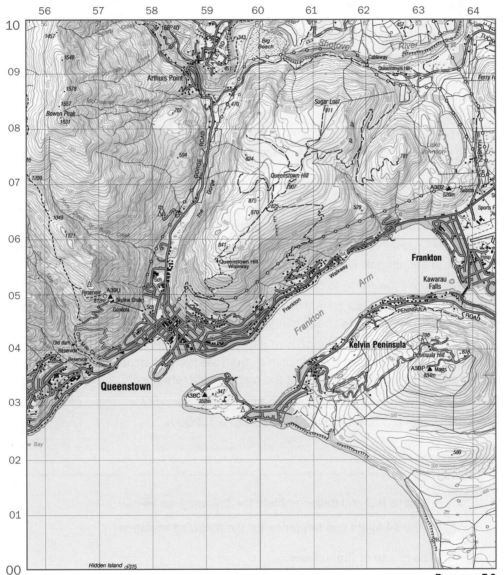

Resource 7.3

2   Study the map above and then complete the following task.

   a   Identify the feature located at the following grid references:

      i   GR 639063 _____   ii   GR 572049 _____

      iii GR 638053 _____   iv   GR 564083 _____

      v   GR 565041 _____   vi   GR 575001 _____

Using the same premise as the six-figure grid reference, the 14-figure grid reference offers an even more precise method of finding absolute location. It achieves this by employing seven figures for the easting and northing instead of the three used by the six-figure grid reference system. In doing so it enables the map reader to pinpoint the location of features on a map to an accuracy of ±100 metres.

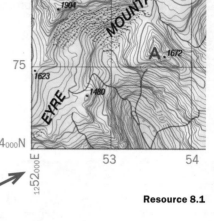

Resource 8.1

To give the absolute location of a feature using 14-figure grid references, follow the steps below:

1 To calculate the easting, record the first two digits of the easting found in the bottom left corner of the map (Resource 8.1). The first two figures of the easting are *12*.

2 Identify the first vertical grid line to the left of the symbol or feature you are finding the absolute location for. For the spot height feature located at **A** (above), the third and fourth figures of the reference are *5* and *3*.

3 Estimate the number of tenths eastward the symbol is from the easting on the left. In the example above, the symbol is located a distance of $\frac{7}{10}$ from the easting on the left. This gives a reading of *7*. This is the fifth number of the easting grid reference.

4 The final two numbers of the full easting grid reference are *00* as we can only estimate the reference to the nearest 100 metres.

5 Repeat the above four steps to calculate the northing. The first two figures of the northing are *49*.

6 Now identify the first horizontal grid line below the symbol. In the example above, the first grid line below is labelled *75* in the left-hand margin. These are the third and fourth figures of the northing.

7 Estimate the number of tenths northward the symbol is from the northing beneath. In the example above, the symbol is located a distance of $\frac{2}{10}$ from the northing beneath. This gives a northing reading of *2*. This is the fifth number of the northing grid reference.

8 The final two numbers of the full northing grid reference are *00*.

9 Express the final grid reference as: 1253700 E 4975200 N

Learning Activities

1 Study Resource 8.2 and then complete the following activities.

   a State the 14-figure grid reference for the following features:

      i The mouth of Otahu River _____

      ii Tunaiti Mountain _____

      iii Te Karaka Point _____

      iv Rawengaiti Island _____

      v The Waikiekie Road Quarry _____

      vi Substation south of Wentworth Valley Road _____

b Identify the feature located at the following grid references:

i 1853100 E 5878700 N _____

ii 1853900 E 5875600 N _____

iii 1855500 E 5880800 N _____

iv 1855600 E 5877500 N _____

v 1856600 E 5877500 N _____

vi 1853000 E 5875800 N _____

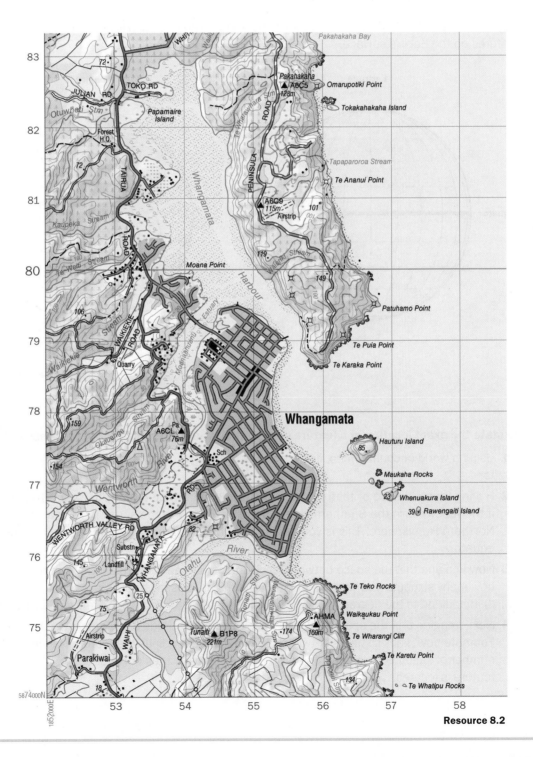

Resource 8.2

Another useful method of determining absolute location is to use latitude and longitude.

Lines of latitude are imaginary lines that run east to west around the Earth's circumference. They run parallel to each other and for this reason are also known as *parallels of latitude*. Latitude is measured in degrees north (N) or south (S) in relation to the equator, which itself represents zero degrees (0˚). The *equator* divides the earth into the Northern Hemisphere and Southern Hemisphere. The latitude of the North Pole is 90˚N of the equator while the latitude of the South Pole is 90˚S of the equator.

Lines of longitude run in a north to south direction from the North Pole to the South Pole and on a two-dimensional map, intersect lines of latitude at right angles. Longitude is measured in degrees west (W) or east (E) of the prime meridian (0˚). The *prime meridian* divides the earth into the Western Hemisphere and Eastern Hemisphere.

Together, lines of latitude and longitude form a grid system that allows us to pinpoint places and features on the Earth's surface.

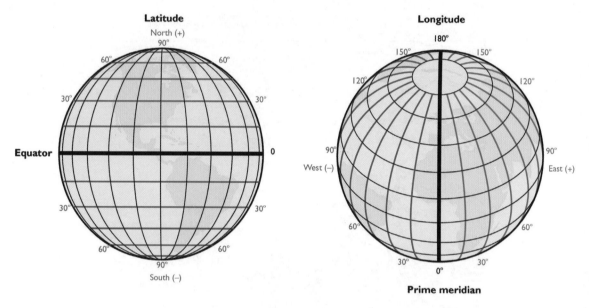

Resource 9.1 Latitude and longitude

To state the exact location of a feature using latitude and longitude, follow the steps below:

1 Using a world or country map, find a place or feature you wish to calculate latitude and longitude for.

2 Having found a place or feature, identify the line of latitude directly to the north of the feature if the feature is in the Southern Hemisphere or directly to the south if the feature is in the Northern Hemisphere. This is your line of latitude in degrees (°) north (N) or south (S) of the equator.

3 Now imagine that the space between each line of latitude is divided into 60 minutes. Estimate how many sixtieths from the line of latitude the feature is located. This is the number of minutes (') in addition to the degrees north or south of the equator your line of latitude is located.

4 Next, imagine that the space between each minute of latitude is divided into 60 seconds. Estimate how many sixtieths between the two degree readings the feature is located. This is the number of seconds (") in addition to the degrees and minutes north or south of the equator your feature is located.

5 Repeat the process above to identify the line of longitude. If your feature is located in the Eastern Hemisphere, identify the line of longitude directly to the west. If it is in the Western Hemisphere, identify the line of longitude directly to the east. This is your line of longitude in degrees (°) west (W) or east (E) of the prime meridian.

6 Now estimate the number of minutes and seconds as above.

7 Remember to always state latitude first and longitude second.

To increase accuracy even further, each line of latitude and longitude can be divided into smaller units called minutes and each minute of each degree can be divided into seconds. There are 60 minutes in each degree of latitude or longitude and 60 seconds in every minute.

Topography and bathometry of the Barents Region

Lines of longitude identify the direction of north.

Lines of longitude run north-south from one pole to the other.

Arctic circle 70°N.

The 'N' identifies the region as being located in the Northern Hemisphere.

Lines of latitude run parallel to each other.

The 'E' indicates the region is located in the Eastern Hemisphere.

Resource 9.2

For example, Resource 9.2 shows that the Barents Region, which covers the area of western Russia and the northern areas of Finland, Sweden and Norway, is bounded by the latitudes 60°N and 70°N and the longitudes 15°E and 60°E. However, to give a precise reading of an exact location requires use of degrees and minutes. For example, the exact location of Stockholm, Sweden could be expressed as 57° 17'N 18° 03'E, which means it is 57 degrees and 17 minutes north of the equator and 18 degrees and 03 minutes east of the prime meridian.

1 With the aid of an atlas, complete the following table.

Belém, Brazil	
	12° 00' S / 77° 02' W
	35° 45' N / 51° 45' E
Cairo, Egypt	
Beijing, China	
	32° 57' N / 13° 12' E
	41° 17' S / 174° 47' E
Suva, Fiji	
Cape Town, South Africa	
	64° 04' N / 21° 58' W

2 Using an atlas and the map in Resource 9.3, estimate the latitude and longitude of the following places using degrees, minutes and seconds.

a Whangarei _____

b Hamilton _____

c Napier _____

d Hastings _____

e Palmerston North _____

f Wellington _____

g Nelson _____

h Timaru _____

3 Using an atlas and the map in Resource 9.3, identify the name of the city located at:

a 45° 1' 48" S / 168° 39' 46" E _____

b 46° 24' 0" S / 168° 21' 0" E _____

c 39° 56' 0" S / 175° 3' 0" E _____

d 38° 8' 16" S / 176° 15' 5" E _____

e 41° 17' 0" S / 173° 17' 0" E _____

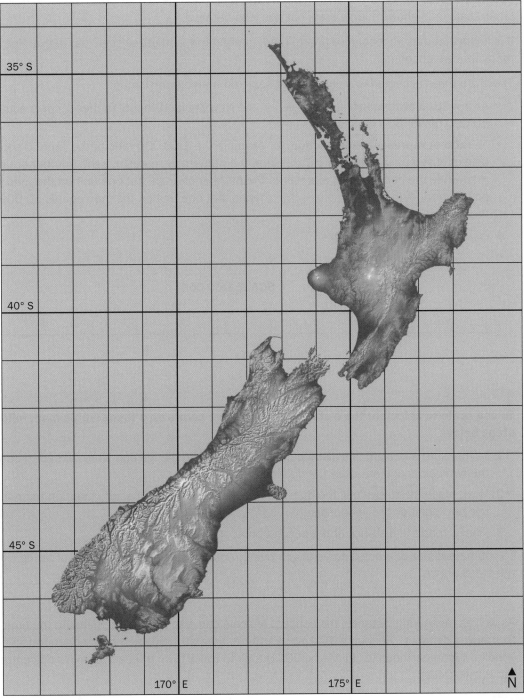

Resource 9.3

A map represents an area on the Earth's surface and a map's scale refers to the relationship between distances portrayed on a map and the distance on the ground. Put in another way, a map is a scaled down representation of part of the Earth's surface. This means that a map's scale can be used to calculate surface distances represented by the map.

A map which depicts a small area of land is referred to as a large scale map. This is because the surface area represented by the map has only been scaled down a small proportion, meaning, the scale is larger. A large scale map shows a small area, but in great detail. On the other hand, a map depicting a large area, such as an entire country or continent, is considered a small scale map. This is because in order to show the entire country the map must be scaled down until it is much smaller. A small scale map shows a large area, but it is less detailed.

It is important that we recognise how to read, understand, and utilise scale as we examine the various maps that we encounter.

There are three accepted ways of showing scale on a topographic map:

- As a **written statement**, for example, on the New Zealand Topo50 series 1 cm is equal to 50 000 cm (or 500 metres).

- As **ratio or representative fraction**, for example the 1:50 000 ratio of the Topo50 series can be expressed as the fraction $\frac{1}{50\,000}$. Here, the numerator represents the number of units on the map while the denominator represents the number of units that one unit on the map is equal to on the ground. In this example, $\frac{1}{50\,000}$ means that one unit on the map equals 50 000 units on the ground.

- As a **linear scale** or scale bar:

SCALE 1:50 000

To use scale to calculate the distance between two points on a topographic map, follow the steps below:

1 To measure a straight line distance, place a ruler (or the edge of a sheet of paper) between the two points and measure the distance between them.

2 Next, place the ruler along the map's linear scale, overlaying the zero point on the ruler with the zero point on the linear scale.

3 Finally, read the distance of the second point off the linear scale.

To estimate distance along a curve (e.g. a river), replace the ruler with a piece of string and follow the steps above.

As well as measuring distance, the concept of scale can also be used to measure the time it takes to travel a distance. Resource 10.1 shows the estimated time the tsunami generated from the Tohoku, Japan earthquake (11 March 2011) took to travel from the earthquake's epicentre across the Pacific Ocean.

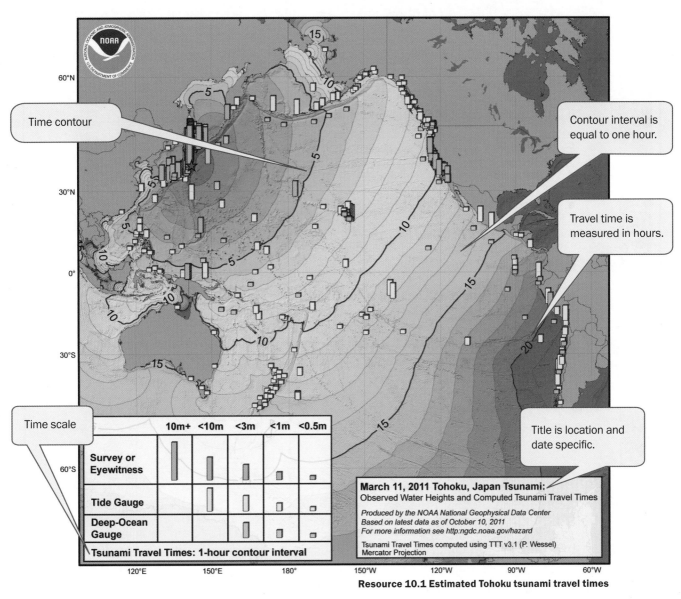

Resource 10.1 Estimated Tohoku tsunami travel times

1 Express the following scales as statements:

a 1:2 million _____

b 1:250 000 _____

c 1:50 000 _____

d $\frac{1}{25}$ _____

e $\frac{1}{2500}$ _____

2 Express the following scales as representative fractions and ratios:

a 1 cm is equal to 5000 m _____

b 1 cm is equal to 200 m _____

c 1 cm is equal to 10 km _____

d 1 cm is equal to 20 km _____

3 Study Resource 10.1 and calculate how many hours it took the Tohoku tsunami to reach:

a South Korea _____

b The eastern coast of Queensland, Australia _____

c Fiji Islands _____

d Alaska, USA _____

e The coast of New South Wales, Australia _____

f The west coast of Mexico _____

g The coast of Chile _____

h The northern tip (Cape Reinga) of New Zealand _____

4 With reference to Resource 10.1, explain how time-scale maps can assist with civil emergency management during a hazard event.

_____

_____

_____

_____

_____

_____

_____

5 Study Resource 10.2 and measure the straight line distance:

a Between Captain Cook's Anchorage located at GR 069999 and Tapeka Point

_____

b Of Long Beach in Russell _____

c Between Tapeka Point and Hermione Rock _____

d Of the route taken by the Opua vehicular ferry _____

e Of the route taken by the Paihia to Russell vehicular ferry _____

f Of the aerial cable crossing Paroa Bay _____

g Of Toretore Island at its widest point _____

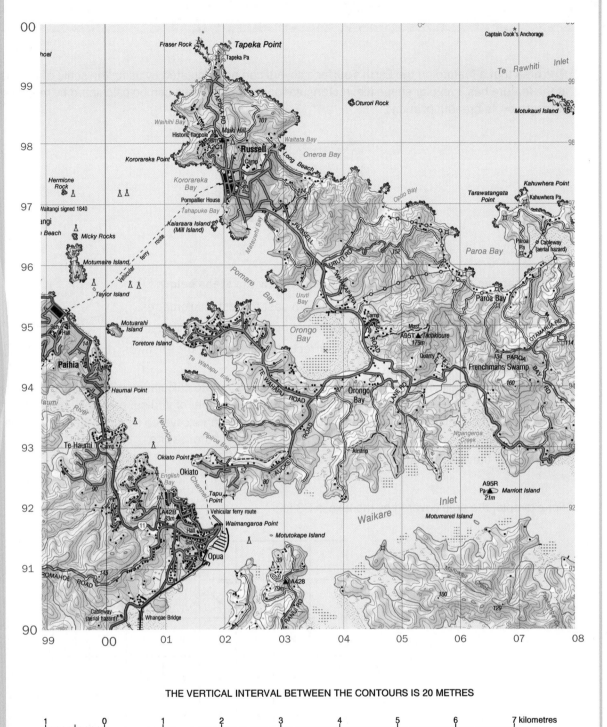

THE VERTICAL INTERVAL BETWEEN THE CONTOURS IS 20 METRES

Resource 10.2 Topographic map of the Bay of Islands

The area that a feature on the Earth's surface covers can be calculated by using the scale of the map. If the feature has a regular shape (i.e. rectangular or square), its area can be calculated by multiplying it length by its breadth (x and y).

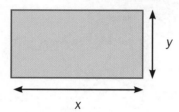

To estimate an area from a topographic map, follow the steps below:

1 Measure the distance of the map area from east to west in kilometres.

2 Measure the distance of the map area from north to south in kilometres.

3 If the area to be measured is square or rectangular, multiply the east-west distance by the north-south distance.

4 If the area to be measured is an irregular shape, an accurate calculation will be difficult. Instead, an estimate of the area can be made by counting the number of grid squares covered by the feature. This can be done by counting the number of squares covered by more than half of the feature and ignoring squares covered by less than half of the feature.

5 Express the area in square kilometres (km^2).

For example, the area covered by the lake is 58 km^2.

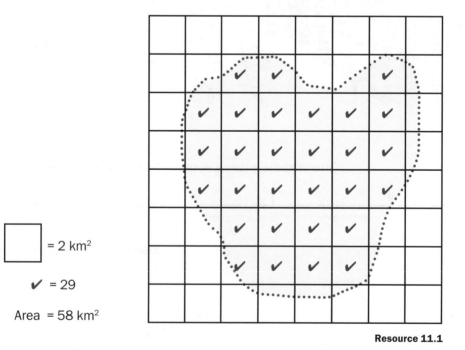

$\square$ = 2 km^2

✔ = 29

Area = 58 km^2

Resource 11.1

Geography Skills for NCEA Level Two

ISBN: 9780170389341

1 Study Resource 11.2 and then complete the following activities.

a Calculate the area bounded by the map. _____

b Estimate the area of Lake Rotongaro. _____

c Estimate the area of Lake Waikare. _____

d Estimate the area in the map covered by native forest (   ). _____

e Estimate the area covered by the flow of the Waikato River and its flood banks.

_____

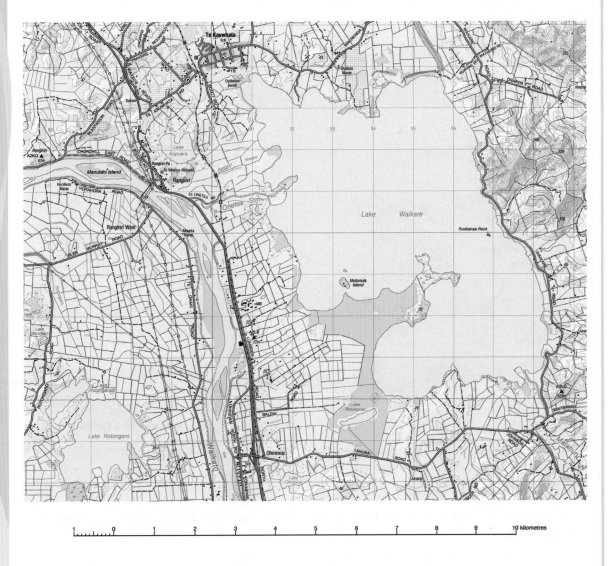

Resource 11.2 Topographic map of Lake Waikare, Waikato

Relief is a word used by geographers to describe the shape or pattern of landforms, its height above sea level and the steepness of its slopes.

Topographic maps show the pattern of relief in three ways:

- shading
- spot heights
- contour lines.

Of the three methods used, contour lines are most effective in showing relief patterns. *Contour* lines trace out areas of equal height or elevation above sea level, and give an indication of the *gradient* or slope of the land. Closely spaced contour lines indicate a steep gradient and contour lines spaced far apart indicate a gentle gradient (Resource 12.1).

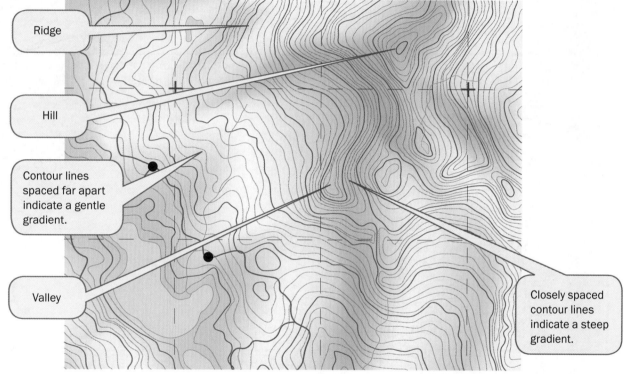

Ridge

Hill

Contour lines spaced far apart indicate a gentle gradient.

Valley

Closely spaced contour lines indicate a steep gradient.

Resource 12.1 Contour map

The vertical distance between adjacent contour lines is known as the contour interval (CI). The interval is always constant on any given map.

To use contour lines to determine the elevation of a feature on a topographic map, follow the steps below:

1 Find the contour interval of the map from the key or legend, and note both the interval and the unit of measure. New Zealand's Topo50 map series, for example, has a contour interval of 20 metres.

2 Find the numbered contour line nearest the feature for which the elevation is being sought.

3 Determine the direction of slope from the numbered contour line to the desired feature.

4 Count the number of contour lines that must be crossed to go from the numbered line to the feature and note the direction up or down. The number of lines crossed multiplied by the contour interval is the distance above or below the starting value.

5 When the feature is on a contour line, its elevation is that of the contour. If the feature is between contour lines, then estimate the elevation to be one-half of the contour interval.

To construct a cross-section, follow the steps below:

1 Decide where you want the cross-section line to be.

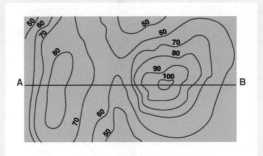

2 Place the edge of a sheet of paper along the line that joins points *A* and *B*. Mark points *A* and *B* on the edge of the sheet of paper.

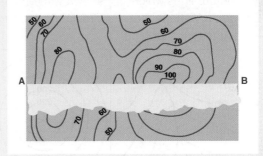

3 Beginning at point *A*, mark the position of each contour line that touches the edge of the sheet of paper between point *A* and *B*. Write the value of each contour line adjacent to each mark. Also mark the location of any major feature such as a road or river.

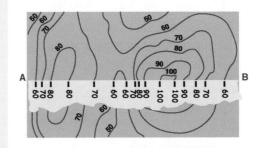

4 Draw the horizontal and vertical axes for your cross-section. The horizontal axis should be drawn the same length as the line *A–B* on the contour map. The vertical axis should be scaled to reflect the contour interval of the map (e.g. 0, 10, 20, 30 and so on).

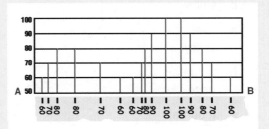

5 Place the sheet of paper along the horizontal axis and plot the contour elevations as though you were plotting a line graph.

Join the dots with a smooth curved line. Draw arrows to show the location of any important features.

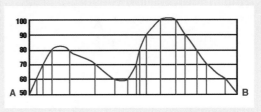

Resource 12.2

ISBN: 9780170389341

Common contour patterns

A skilled geographer can visualise the shape of the landforms by studying patterns created by contour lines.

Landform	A cliff or steep gradient has contour lines that are very close together	Gentle gradients have widely spaced contour lines	A valley has contour lines which form a U shape that points to a higher land
Contour Pattern	120 100 80 60 40 20	20 40 60	60 80 100 120

Landform	A river valley has contour lines which form a V shape that points to higher land	Round hills are shown with enclosed circular contour lines	Enclosed contour lines that have a linear shape indicate a mountain ridge	Lines jutting away from a hill or mountain indicate a spur
Contour Pattern	160 140 120 100	180 200 220 ▲	80 100 120 120 100 80 60	80 100 120 160 160

Resource 12.3

1 Using Resource 12.3 as a guide, locate and label the contour patterns for the following features in the map below.

 a steep gradient

 b gentle gradient

 c round hill

 d valley

 e river valley

 f mountain ridge

 g mountain spur

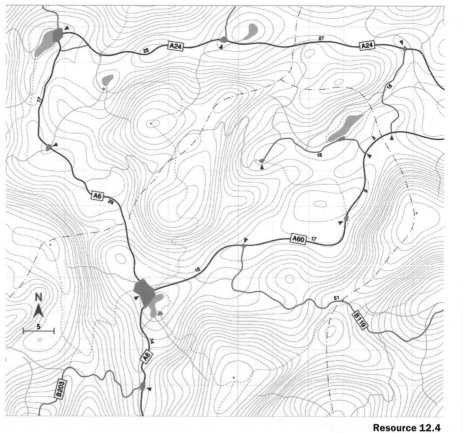

Resource 12.4

Geography Skills for NCEA Level Two
ISBN: 9780170389341

2 Construct a cross-section of Frankton Arm, Queenstown from Point A to Point B using a vertical contour interval of 20 m.

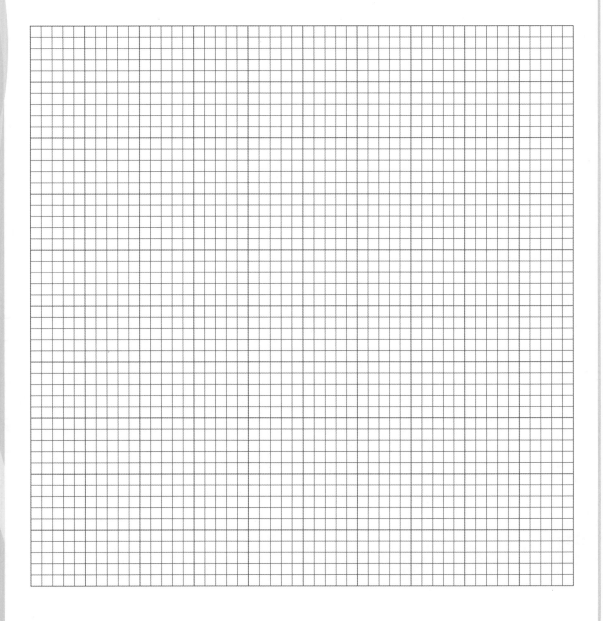

3 Construct a cross-section of Perseverance Harbour, Campbell Island from Point A to Point B using a vertical contour interval of 20 m.

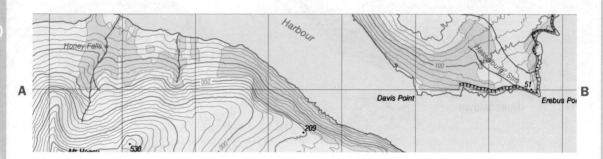

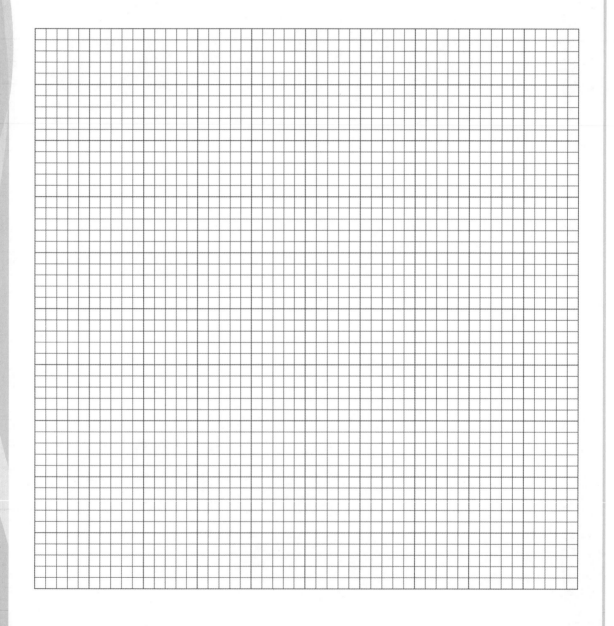

Topographic maps contain an immense amount of detail so it is sometimes helpful to construct a précis map to highlight just a few of the features illustrated on the map or photograph. Précis maps are another useful tool for geographers as they enable the relationship between features to be identified easily.

To construct a précis map, follow the step below:

1 Identify from the topographic map the particular feature or features you wish to study.

2 Establish the relative location of the feature on the topographic map (or absolute location if the use of a six-figure grid reference is required).

3 Draw a simple outline of each feature on your précis map, taking care to place features in their correct location relative to others. You may choose to draw a grid on your précis that corresponds to the grid on the topographic map to ensure even greater accuracy. Shade in each feature in an appropriate colour, for example, blue for water features, black for cultural features and green for vegetation.

4 Summarise important features by constructing a key or legend to show the meaning of any symbols or shading used.

Land use surrounding Lakes Waikare and Rotongaro

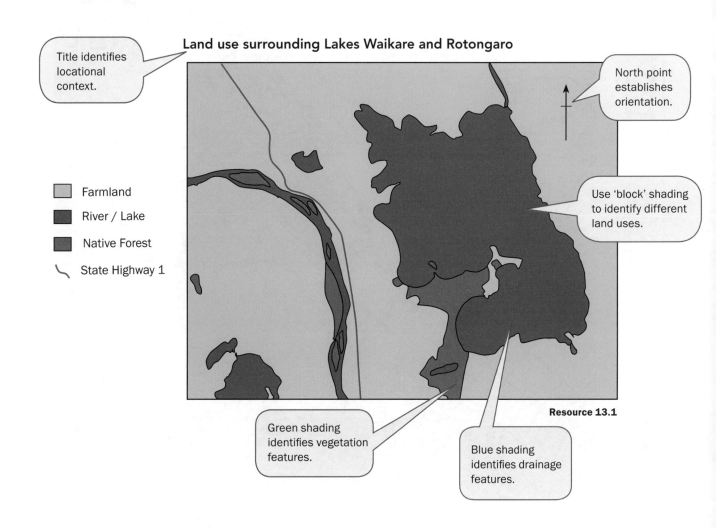

Title identifies locational context.

North point establishes orientation.

Use 'block' shading to identify different land uses.

Green shading identifies vegetation features.

Blue shading identifies drainage features.

Farmland
River / Lake
Native Forest
State Highway 1

Resource 13.1

1 Study the topographic map of the Richmond-Nelson region (Resource 13.2). Construct a précis map that includes the following features:

a Coastline

b Richmond and Nelson urban area

c Rabbit Island

d Mountainous area to the southeast

e State Highway 6

2 Construct a key to summarise the features in Question **1**.

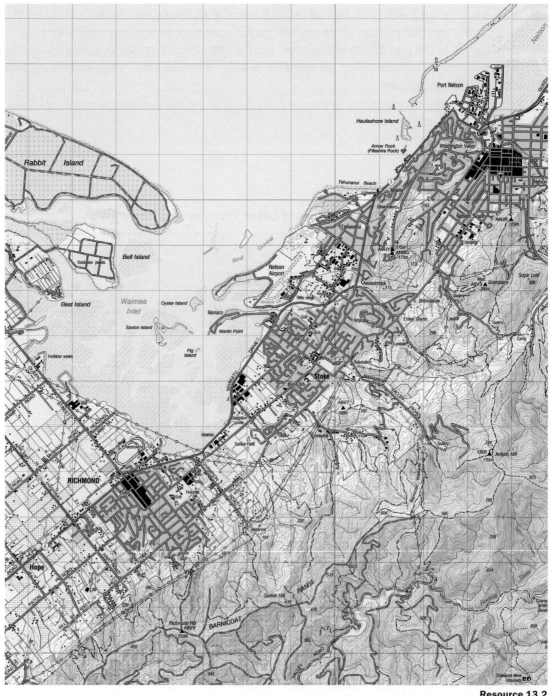

Resource 13.2

Key:

☐ Coastline

☐ Richmond and Nelson urban area

☐ Rabbit Island

☐ Mountainous area to the southeast

☐ State Highway 6

3 Study the topographic map of Gisborne-Poverty Bay (Resource 13.3). Construct a précis map that includes the following features:

  a Coastline

  b Gisborne urban area

  c Railway line

  d Turanganui and Taruheru Rivers

  e The natural coastal feature located at Tuaheni Point

4 Construct a key to summarise the features in Question 3.

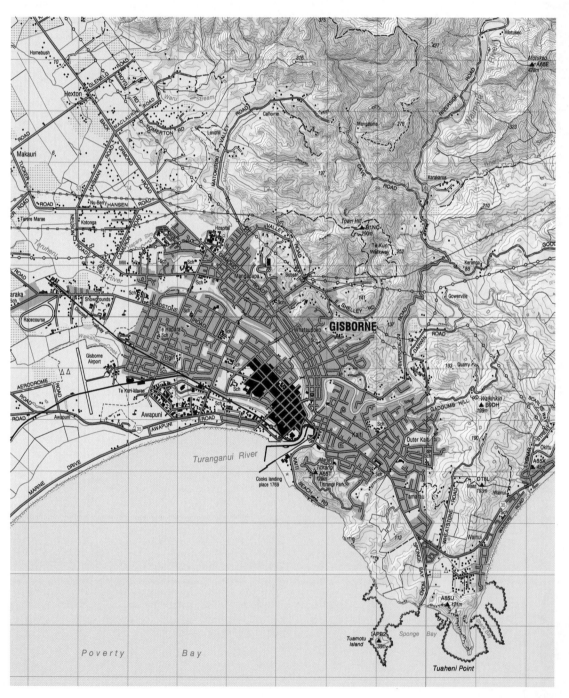

Resource 13.3

Key:

☐ Coastline

☐ Gisborne urban area

☐ Railway line

☐ Turanganui and Taruheru Rivers

☐ The natural coastal feature located at Tuaheni Point

Graph interpretation and construction

Graphs offer geographers a simple and effective way of presenting statistical information. For example, geographers routinely use graphs to:

- Compare two or more variables
- Illustrate change over time
- Illustrate the relationship between two sets of data
- Identify patterns or trends
- Show how something is made up.

There are many types of graph, each designed to present information in a specific way. Since each graph type is suited for illustrating different types of information, it is important you learn how to interpret and construct a variety of graph types. In this chapter, you will learn how to interpret and construct bar, line, pie and percentage bar graphs, scatter graphs, triangular graphs, pictograms, age-sex pyramids and climate graphs.

In Level 2 NCEA Geography, you will on occasion be given geographic data and asked to construct a graph using an appropriate graphing technique of your choosing. Regardless of the type of graph you construct to present data, it is important that you follow accepted graphing conventions.

Selection of a suitable graphical technique

Each characteristic of the data can only be suitably represented by an appropriate graphical technique. For example, to show continuous data such as that related to the temperature or growth of population, line graphs are used. Similarly, bar graphs are frequently used to show rainfall or the production of commodities.

Selection of a suitable scale

A graph is always drawn to a scale. The scale must cover the entire data that is to be represented. Therefore, the scale should neither be too large nor too small. Ensure both axes are long enough to accommodate the range of data you wish to show, as the use of broken axes is generally unacceptable.

Design

The title of the graph must be clear and include:

- The name of the area
- Reference year of the data used
- The variable plotted in the graph.

The title, subtitle and the corresponding year is shown in the centre at the top of the graph.

Legend or key

The legend (or key) must clearly explain the colours, shades and symbols used in the graph. It is usually positioned at the lower left or lower right side of graph.

14 | Bar graphs (single and multiple)

Bar graphs are the simplest way to compare two sets of information. Generally, bar graphs consist of horizontal bars while column graphs use vertical bars. However, for the purposes of NCEA Geography you will not usually be required to make a distinction between horizontal bar and vertical column graphs.

To construct a simple bar graph, follow the steps below:

1 Decide what information is to be plotted on each axis. In most cases, you will plot the non-quantifiable variable (i.e. the one that does not change) on the x-axis (e.g. country names, age groups, or periods such as months or years) while the quantifiable data is normally plotted on the y-axis. It is for this reason the y-axis is sometimes referred to as the variable axis.

2 Bar graphs abide by the graphing convention that requires the y-axis (variable axis) to follow a constant number scale starting from zero (e.g. 0, 5, 10, 15 or 0, 10, 20, 30). You will therefore need to determine an appropriate scale for the variable axis.

3 Having determined the range and scale of the data to be plotted, use a ruler to construct the axes. Like most graphs, bar graphs have two axes: the x-axis is usually horizontal (i.e. runs across the bottom of the graph) while the y-axis is usually vertical (i.e. runs up the left-hand side of the graph). Ensure both axes are long enough to accommodate the range of data you wish to show.

4 Label each axis (including the units of measurement) and give the graph a title that clearly states what the graph illustrates. If appropriate, the title should also include location and date specific information.

5 Use a ruler to construct each bar. Ensure all bars are drawn with constant spacing and equal width.

6 Shade in each bar with colour pencil. If appropriate, label each bar or include a key if you are constructing a multiple bar graph (Resource 14.2).

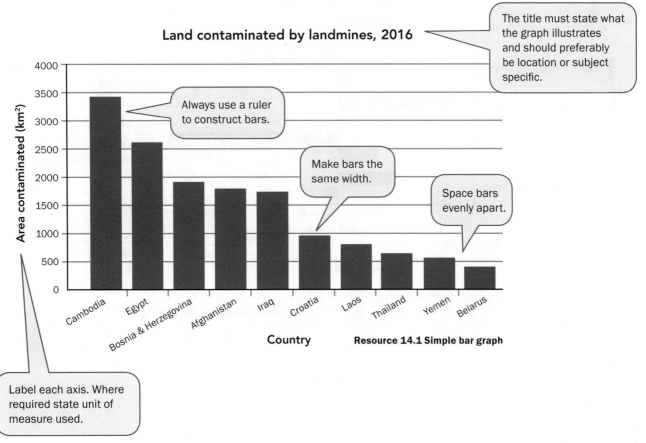

Resource 14.1 Simple bar graph

Population of the largest countries in Latin America and the Caribbean, 1980 and 2015

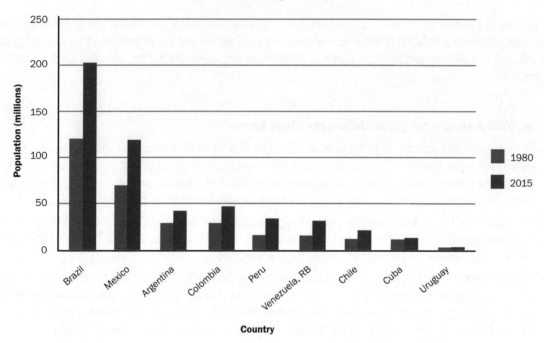

Resource 14.2 Multiple bar graph

The title must state what the graph illustrates and should preferably be location and date specific.

Net migration by continental region, 1960 and 2010

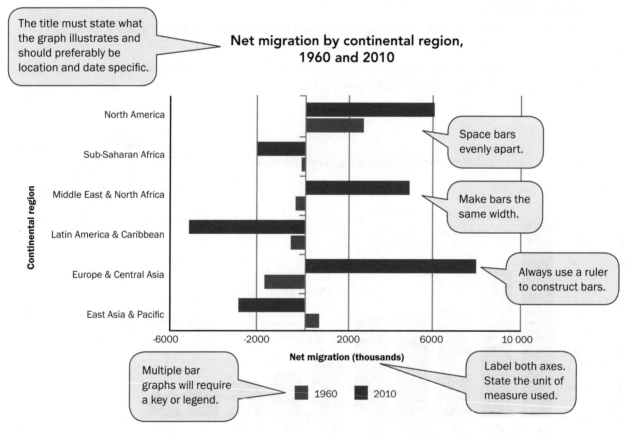

Space bars evenly apart.

Make bars the same width.

Always use a ruler to construct bars.

Multiple bar graphs will require a key or legend.

Label both axes. State the unit of measure used.

Resource 14.3 Multiple bar graph

1 Study Resource 14.1 and then complete the following activities.

   a State the land area in Iraq contaminated with landmines. _____

   b Estimate the number by which the area of landmine contamination in Cambodia exceeded that of Yemen in 2016. _____

2 Study Resource 14.3 and then complete the following activities.

   a Name the continental region that experienced the largest net migration increase in:

     i 1960 _____ ii 2010 _____

   b Name the continental region that experienced the largest net migration decrease in:

     i 1960 _____ ii 2010 _____

   c State the net migration for Sub-Saharan Africa in:

     i 1960 _____ ii 2010 _____

   d Calculate the increase in net migration in North America between 1960 and 2010.

     _____

   e With reference to Resource 14.3, describe the changes in migration flows between 1960 and 2010.

     _____

     _____

     _____

     _____

     _____

     _____

   f Suggest a reason for the change in migration flows from 1960 to 2010.

     _____

     _____

     _____

     _____

     _____

     _____

     _____

     _____

3 Use the data in Table 14.1 to construct a bar graph showing the 10 top most pirated movies of 2015.

Movie	Total downloads 2015 (millions)
Interstellar	46.8
Furious 7	44.8
Avengers: Age of Ultron	41.6
Jurassic World	36.9
Mad Max: Fury Road	36.4
American Sniper	34.0
Fifty Shades of Grey	32.1
The Hobbit: The Battle of the Five Armies	31.6
Terminator: Genisys	31.0
Kingsman: The Secret Service	30.9

Table 14.1

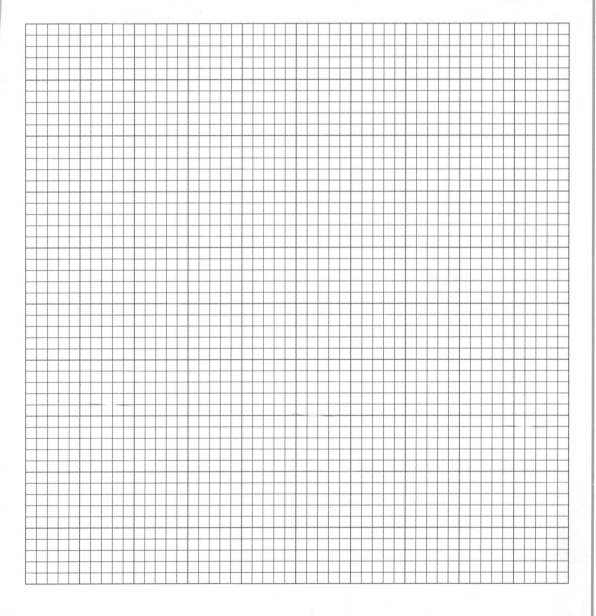

4 Use the data in Table 14.2 to construct a multiple bar graph showing the percentage of foreign born in Australia, Canada and the United States by region of birth.

	Australia (%)	Canada (%)	USA (%)
Africa	5.6	6.1	4.0
Americas	4.1	15.3	55.1
Asia	31.8	40.8	28.2
Europe	47.1	36.8	12.1
Oceania	11.2	0.9	0.5

Table 14.2

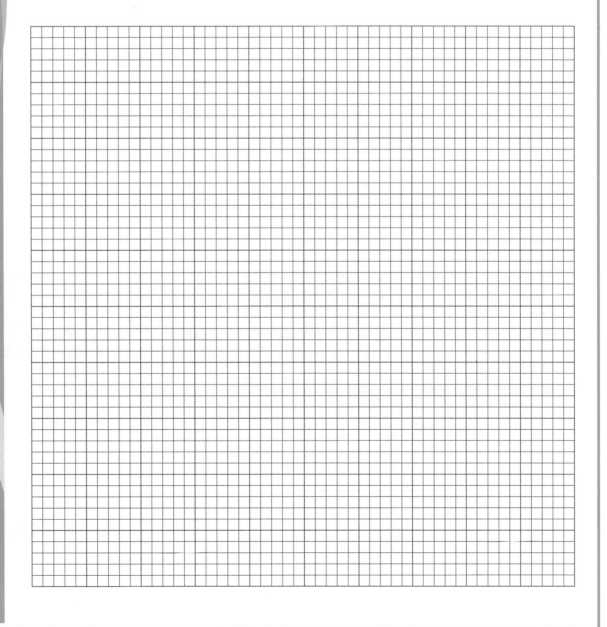

Line graphs are especially useful in geography as they are easy to create, and their visual characteristics reveal data trends clearly.

Like bar graphs, line graphs provide a visual representation of how two variables (shown on the *x*-axis and *y*-axis) relate. The vertical *y*-axis in a line graph usually indicates quantity (e.g. population size or change, volume) or percentage in the case of a compound graph, while the horizontal *x*-axis often measures units of time (e.g. years). As a result, the line graph is often used to show change over time. For example, if you wanted to graph changes in the crude birth rate over time, you could measure the time variable in years along the *x*-axis, and birth rate (per 1000) along the *y*-axis (Resource 15.1).

Multiple line graphs differ from single line graphs in that they plot two or more related sets of data on the same graph, allowing for easy data comparison (Resource 15.2).

Not all graphs show information from the past. Some are used to show estimates about the future. These estimates are called projections (that is, what we predict will happen in the future, based on our analysis of past and current trends). For example, the line graph in Resource 15.1 shows that New Zealand's crude birth rate is expected to steadily decline in the period leading to 2044. This projection was determined by analysing current trends in New Zealand's birth rate and then by projecting those trends into the near future.

Projected data is usually shown on a line graph by a dotted or dashed line.

To construct a simple line graph, follow the steps below:

1 Decide what information is to be plotted on each axis. In most cases, you will plot the non-quantifiable variable (i.e. the one that does not change) on the *x*-axis (e.g. country names, age groups, or periods such as months or years) while the quantifiable data is normally plotted on the *y*-axis. It is for this reason the *y*-axis is sometimes referred to as the variable axis.

2 Like bar graphs, line graphs also abide by the graphing convention that requires the axes to follow a constant number scale starting from zero (e.g. 0, 5, 10, 15 or 0, 10, 20, 30). You will therefore need to determine an appropriate scale for the variable axis.

3 Having determined the range and scale of the data to be plotted, use a ruler to construct the axes. Like most graphs, simple line graphs have two axes: the *x*-axis is usually horizontal (i.e. runs across the bottom of the graph) while the *y*-axis is usually vertical (i.e. runs up the left-hand side of the graph). Ensure both axes are long enough to accommodate the range of data you wish to show.

4 Label each axis (including the units of measurement) and give the graph a title that clearly states what the graph illustrates. If appropriate, the title should also include location and date specific information.

5 Next, plot each value on the graph, and then join the points together with either a ruler for a straight-line curve or freehand if a smooth curve is required.

6 If you are constructing a multiple line graph, it is recommended that you also include a key (Resource 15.2).

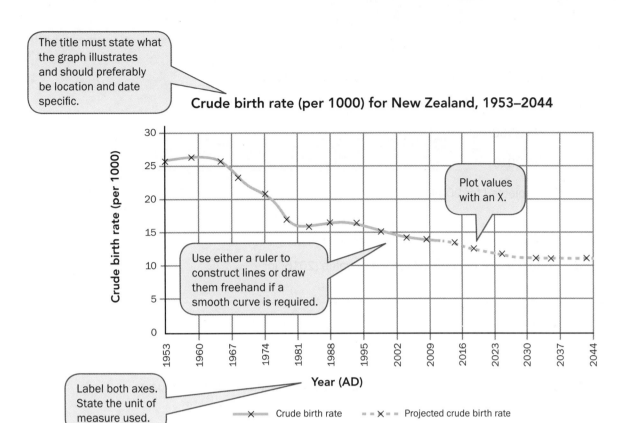

Crude birth rate (per 1000) for New Zealand, 1953–2044

Resource 15.1

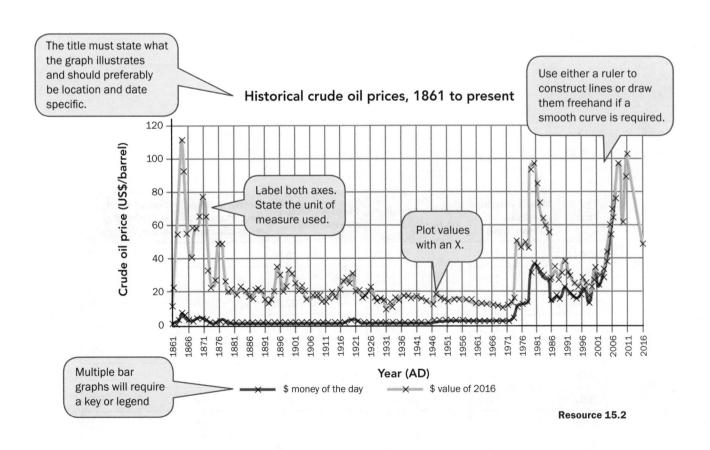

Historical crude oil prices, 1861 to present

Resource 15.2

1 Study Resource 15.1 and then complete the following activities.

    a State the crude birth rate for New Zealand:

        i In 1960 _____

        ii In 2010 _____

        iii Projected for 2044 _____

2 Study Resource 15.2 and then complete the following activities.

    a State the cost of crude oil in today's value (2016) in:

        i 1861 _____

        ii 1864 _____

        iii 1979 _____

    b State the cost of crude oil in the year it was purchased (money of the day):

        i 1861 _____

        ii 1979 _____

        iii 2011 _____

    c Explain why the value of crude oil is the same for both lines in 2016.

    _____

    _____

    _____

3 Use the data in Table 15.1 to construct a multiple line graph to show the number of telephone subscriptions and Internet connections per 100 people worldwide.

Year	Mobile cellular telephone subscriptions	Individuals using the Internet	Fixed-telephone subscriptions	Active mobile-broadband subscriptions	Fixed (wired) broadband subscriptions
2001	15.5	8.0	16.6		0.6
2002	18.4	10.7	17.2		1.0
2003	22.2	12.3	17.8		1.6
2004	27.3	14.1	18.7		2.4
2005	33.9	15.8	19.1		3.4
2006	41.7	17.6	19.2		4.3
2007	50.6	20.6	18.8	4.0	5.2
2008	59.7	23.1	18.5	6.3	6.1
2009	68.0	25.6	18.4	9.0	6.9

(Table continued over page.)

Year	Mobile cellular telephone subscriptions	Individuals using the Internet	Fixed-telephone subscriptions	Active mobile-broadband subscriptions	Fixed (wired) broadband subscriptions
2010	76.6	29.4	17.8	11.5	7.6
2011	83.8	32.5	17.2	16.7	8.4
2012	88.1	35.5	16.7	21.7	9.0
2013	93.1	37.9	16.2	26.7	9.4
2014	95.5	40.4	15.8	32.0	9.8

Table 15.1

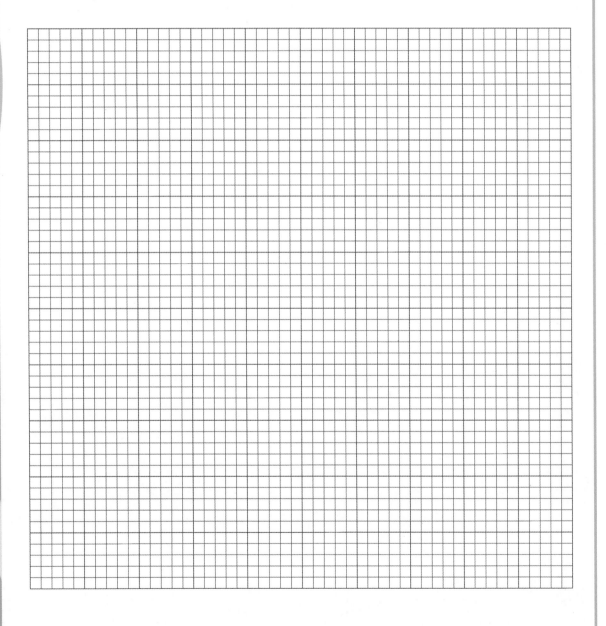

4 Use the data in Table 15.2 to construct a multiple line graph showing population change in selected regions of the world.

Census year	More developed regions (000s)	Less developed regions (000s)	Least developed regions (000s)
1990	1 147 345	4 143 107	524 764
1995	1 174 680	4 538 393	599 098
2000	1 194 967	4 920 400	676 929
2005	1 216 550	5 295 726	761 847
2010	1 237 229	5 671 460	854 697
2015	1 251 351	5 143 963	954 158

Table 15.2

Pie graphs

A pie graph is a useful way of summarising data that has been categorised or represents different values of a given variable (e.g. percentage distribution). A pie graph is circular and divided into a series of segments, each segment representing a particular category. The area of each segment is always the same proportion of the circle as the category is of the total data set. For example, the pie graph in Resource 16.1 clearly shows that 41% of New Zealand's fresh-water supply is allocated to hydro and that only 5% is used for domestic supply.

To construct a pie graph, follow the steps below:

1 Use a compass or stencil to construct a circle. Then draw a line from the centre of the circle to the 12 o'clock position.

2 Convert each percentage value into degrees by multiplying each variable by a factor of 3.6. For example, in Resource 16.1 the irrigation portion of the graph was calculated by multiplying the percentage allocated to irrigation by 3.6 to represent 165.6° (i.e. 46% x 3.6 = 165.6°).

3 Working clockwise from the 12 o'clock position, use a protractor to plot each segment of the pie beginning with largest category followed by the second largest category and so on. If applicable, plot the 'Other' category last.

4 Shade in each segment with coloured pencils. Label each segment or include a key.

5 Give the graph a title that clearly states what the graph illustrates. The title should also include any location and date specific information.

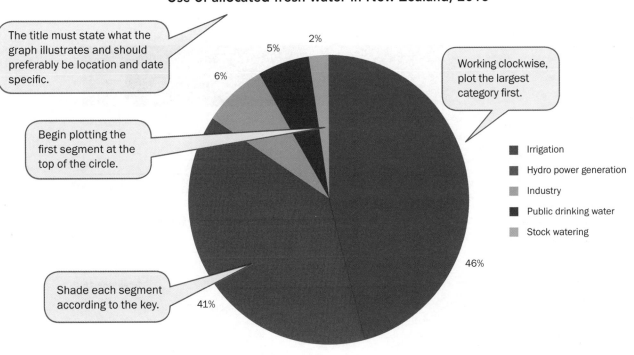

Use of allocated fresh water in New Zealand, 2016

The title must state what the graph illustrates and should preferably be location and date specific.

Working clockwise, plot the largest category first.

Begin plotting the first segment at the top of the circle.

Shade each segment according to the key.

2%
5%
6%
46%
41%

■ Irrigation
■ Hydro power generation
■ Industry
■ Public drinking water
■ Stock watering

Resource 16.1

Percentage bar graphs

Like the pie graph, the percentage bar graph provides a way of summarising data that has been categorised or represents different values of a given variable, however, a percentage bar graph differs to that of a pie graph in that it is constructed in the form of a bar rather than a circle (Resource 16.2). To allow for easy conversion from percentage to segment length, percentage bars are usually drawn 100 mm long.

To construct a percentage bar graph, follow the steps below:

1 Use a ruler to construct a horizontal bar measuring 100 mm by 10 mm.

2 Convert each percentage variable into an equivalent proportional distance. For example, a variable of 25% would be equal to a segment length of 25 mm, and a variable of 14% would be equal to a segment length of 14 mm.

3 Working from the left-hand end of the bar, measure and plot each segment beginning with the largest category followed by the second largest category and so on. If applicable, plot the 'Other' category last.

4 Shade in each segment with coloured pencils. Label each segment or include a key.

5 Give the graph a title that clearly states what the graph illustrates. If appropriate, the title should also include location and date specific information.

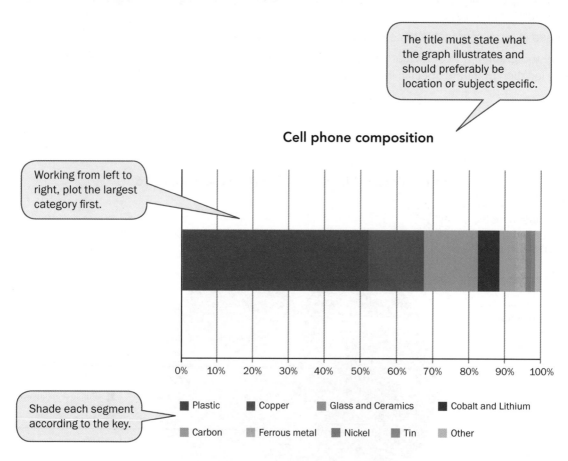

Resource 16.2

1 Use the data in Table 16.1 to construct a pie graph showing the major religions of the world ranked by number of adherents.

Religion	%
Christianity	33.4
Islam	21.0
Hinduism	13.3
Agnosticism	11.8
Chinese Folk Religion	6.6
Buddhism	5.8
Tribal Religion	3.1
Other	5.0

Table 16.1

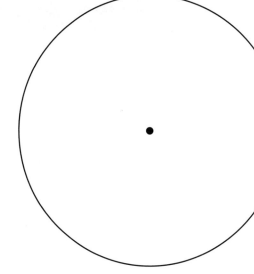

☐ Christianity ☐ Islam ☐ Hinduism

☐ Agnosticism ☐ Chinese Folk Religion

☐ Buddhism ☐ Tribal Religion ☐ Other

2 Use the data in Table 16.2 to construct a pie graph showing the destination of India's electronic exports.

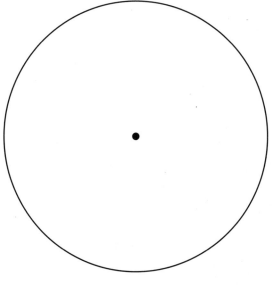

Regions	%
USA	57
Europe	22
South-East Asia	6
Japan	4
West Asia	3
Australia, New Zealand	3
Rest of the world	5

Table 16.2

☐ USA ☐ Europe ☐ South-East Asia ☐ Japan

☐ West Asia ☐ Australia, New Zealand ☐ Rest of the world

ISBN: 9780170389341

3 Use the data in Table 16.3 to construct a percentage bar graph to show greenhouse gas emissions by source.

Greenhouse gas emissions by source (2016)	%
Power generation	24
Land use	18
Agriculture	14
Transport	14
Industry	14
Buildings	8
Waste	3
Other	5

Table 16.3

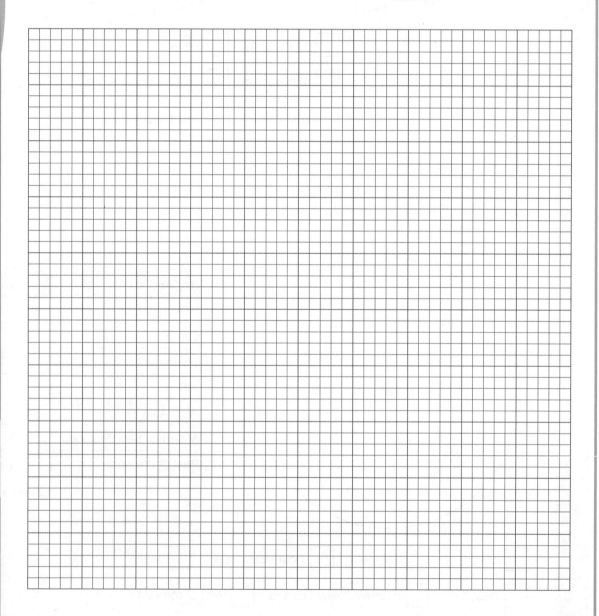

4 Use the data in Table 16.4 to construct two percentage bar graphs showing the frequency of use of print and social media sites.

Frequency of use	Print media (%)	Social media (%)
Less than once a week	34	24
Once a week	34	34
Twice a week	16	16
Every day	16	26

Table 16.4

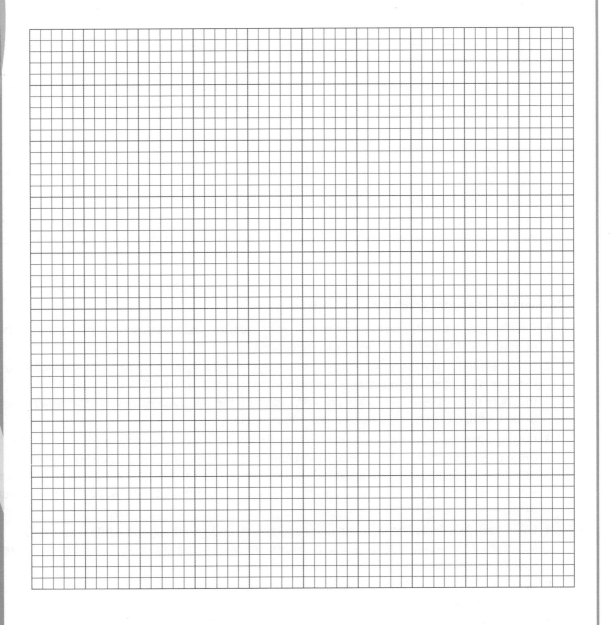

ISBN: 9780170389341

In Geography, a scatter graph is used to present the measurements of two related variables. It is particularly useful when one variable is thought to be dependent upon the values of the other variable.

Data points in a scatter graph are plotted but not joined. The resulting pattern indicates the type and strength of the relationship between the two variables (Resources 17.1 and 17.2).

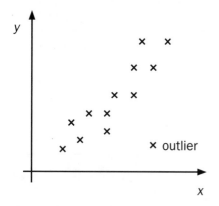

Resource 17.1 An example of a positive correlation

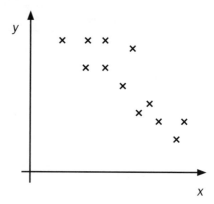

Resource 17.2 An example of a negative correlation

The line of 'best fit' or trend line is a straight line that is drawn on a scatter graph to show the general trend. Lines of best fit help identify the relationship between the two variables shown in the graph. A line of best fit that rises from left to right signifies a positive correlation (i.e. as variable x increases, so too does variable y), whereas a line of best fit that slopes down from left to the right signifies a negative correlation (i.e. as variable x increases, variable y decreases). Data points that are not close to the line of best fit are called outliers.

To construct a scatter graph, follow the steps below:

1 Decide what information is to be plotted on each axis. In most cases, you will plot quantifiable data on both the x-axis and the y-axis (Resource 17.3).

2 Like line and bar graphs, scatter graphs abide by the graphing convention that requires the axes to follow a constant number scale starting from zero (e.g. 0, 5, 10, 15 or 0, 10, 20, 30). You will therefore need to determine an appropriate scale for the two axes. However, if the range of data is too broad, one axis may employ a logarithmic scale.

3 Having determined the range and scale of the data to be plotted, use a ruler to construct the axes. Like most graphs, simple line graphs have two axes: the x-axis is usually horizontal (i.e. runs across the bottom of the graph) while the y-axis is usually vertical (i.e. runs up the left-hand side of the graph). Ensure each axis is long enough to accommodate the range of data you wish to show.

4 Label each axis (including the units of measurement) and give the graph a title that clearly states what the graph illustrates. If appropriate, the title should also include location and date specific information.

5 Next, plot the intersection of each value on the graph with an 'X'. It is important that you do not join the points together.

6 Draw a line of best fit.

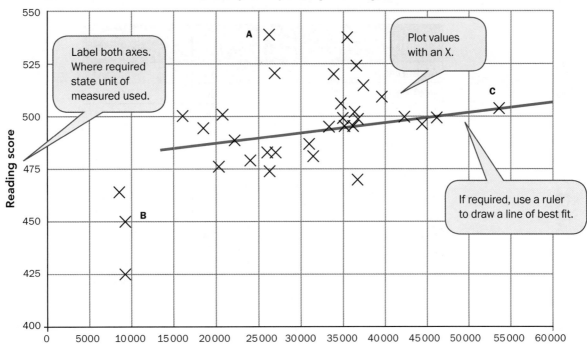

Reading level and national income
(GDP per capita) by country, 2015

The title must state what the graph illustrates and should preferably be location or subject specific.

Label both axes. Where required state unit of measured used.

Plot values with an X.

If required, use a ruler to draw a line of best fit.

Reading score

GDP per capita (USD converted using PPPs)

Resource 17.3

1 Study Resource 17.3 and then complete the following activities.

   a State the reading level and average income per person for:

      i Country A _____

      ii Country B _____

      iii Country C _____

   b Write a statement summarising the relationship between reading score and average GDP per capita.

      _____

      _____

      _____

2 Use the data in Table 17.1 to construct a scatter graph to compare life expectancy with crude birth rates (per 1000).

   a Add a line of best fit to your graph.

   b Write a statement summarising the relationship between life expectancy and crude birth rates.

      _____

      _____

      _____

Country	Crude birth rate (per 1000) 2015	Life expectancy (years) 2015
Afghanistan	34	60
Botswana	25	64
Brazil	15	74
China	12	76
Finland	11	81
France	12	82
India	20	68
Japan	8	84
Malawi	39	63
New Zealand	13	81

Table 17.1

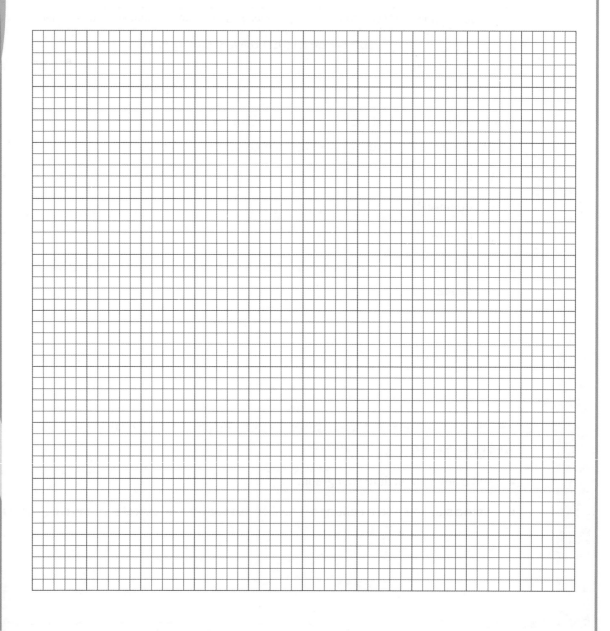

Geography Skills for NCEA Level Two
ISBN: 9780170389341

3 Use the data in Table 17.2 to construct a percentage scatter graph to compare the number of Internet users per 100 people and the percentage of population obesity for selected countries.

a Add a line of best fit to your graph.

b Write a statement summarising the relationship between the number of Internet users per 100 people and obesity for the selected countries.

_____

_____

_____

_____

_____

_____

c Are the two variables necessarily linked (i.e. does a change in one result in a change in the other)?

_____

_____

_____

_____

_____

_____

_____

_____

Country	Obese population (BMI >30) % of total	Internet users per 100 people
Netherlands	10	84
China	5	16
Indonesia	2	6
Italy	21	54
United Kingdom	24	72
Pakistan	5	11
Brazil	11	35
New Zealand	21	70

Table 17.2

A triangular graph (also known as a ternary graph) is a graph that allows geographers to plot and then identify the relationship between three variables. As its name suggests, a triangular graph employs three axes in the form of an *equilateral* triangle (three equal sides), to display the proportions of three related variables (*A*, *B* and *C*) that sum to a constant. For easy analysis, it requires that each variable be expressed as a proportion (or percentage) of the whole.

Added together, a triangular graph's three variables should sum to 100 percent for each individual data sample (i.e. *Variable A* + *Variable B* + *Variable C* = 100%).

In Table 18.1, the three variables that make up each of the 14 samples, sum to 100 percent. For example:

Sample 5: = *Variable A* + *Variable B* + *Variable C*

$\qquad$ = 23.9% + 35.5% + 40.6%

$\qquad$ = 100%

The corresponding location of Sample 5 on a triangular graph is illustrated in Resource 18.1.

The important features of a triangular graph are:

- Each of the three axes is divided into 100 to represent percentages.
- From each axis, lines are drawn at 60° angles to carry the values of each variable.
- The data used must be composed of three related variables, each represented as a percentage value. Accordingly, the sum of the three variables in a triangular graph always adds up to 100.

Sample number	Variable A %	Variable B %	Variable C %	Total %
1	86.9	8.3	4.8	100
2	69.5	21.5	9	100
3	50.4	37.3	12.3	100
4	35.7	45.2	19.1	100
5	23.9	35.5	40.6	100
6	18.8	53.6	27.6	100
7	50	35	15	100
8	42.4	46.6	11	100
9	13.1	57.9	29	100
10	11.8	53.6	34.6	100
11	18.2	48.2	33.6	100
12	10.5	32.4	57.1	100
13	10.1	43.9	46	100
14	10.7	66.3	23	100

Table 18.1

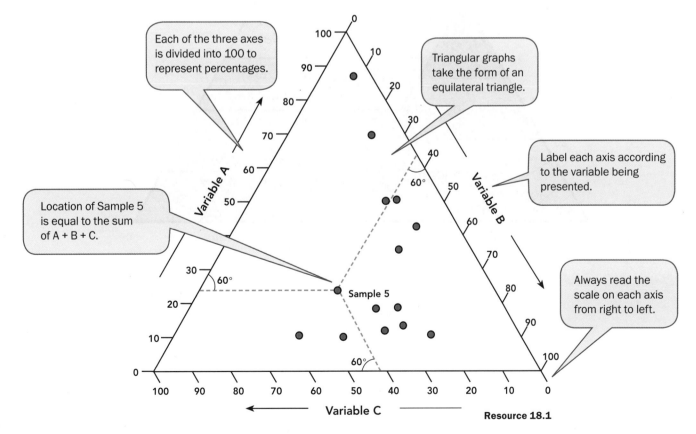

Resource 18.1

To construct a triangular graph, follow the steps below:

1 Draw an equilateral triangle (i.e. a triangle in which all three sides are equal).

2 Divide each side of the triangle into percentages from 0 to 100, so that the upper end of one scale (100%) forms the lower end of the adjoining scale (0%).

3 To plot the position of each data point, it is important that the three related variables for each data set equal 100 (Table 18.1). Each variable will be represented by one side of the triangle. Find the side of the triangle that corresponds with each variable and estimate the percentage each data point is located along each axis. Draw a line from the axis at an angle of 60° and bisect it with the lines generated by the other two variables (Resource 18.1).

It is likely in your study of Level 2 NCEA Geography that you will encounter the following triangular graph of soil texture. Because soil is generally categorised by its composition of three related elements (silt, clay and sand), it is possible to determine the texture classification of a soil sample based on its make up. To use the graph, simply identify the soil sample's silt, sand and clay percentages and follow the colour coded grid line to the point where they intersect. For example, a soil sample comprising 70% silt, 15% clay and 15% sand would be classified silt loam.

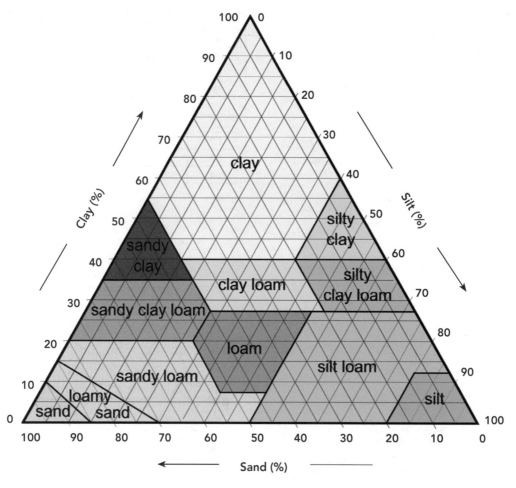

Resource 18.2 Soil texture triangular graph

1 Study Resource 18.2 and then complete the following activity.

    a Identify the texture classification of the following soil samples:

        i Sample 1: 15% silt, 70% clay, 15% sand _____

        ii Sample 2: 40% silt, 35% clay, 25% sand _____

        iii Sample 3: 40% silt, 15% clay, 45% sand _____

2 Use the data in Table 18.2 to construct a triangular graph to compare urban distribution of the states and territories of Australia.

State or territory	Major urban % (>100,000)	Other urban %	Rural %	Total %
Queensland	48	32	20	100
New South Wales	68	21	11	100
ACT	98	0	2	100
Victoria	71	17	12	100
Tasmania	33	42	25	100
South Australia	69	16	15	100
Western Australia	64	20	16	100
Northern Territory	0	66	34	100

Table 18.2

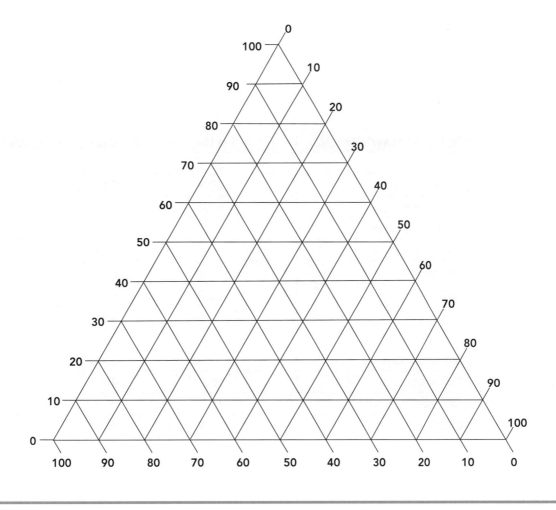

a Describe the urban distribution of Australia's population.

_____

_____

_____

_____

_____

_____

3 Use the data in Table 18.3 to complete the following activities.

a Calculate the values *a* to *i*:

i *a* = _____

ii *b* = _____

iii *c* = _____

iv *d* = _____

v *e* = _____

vi *f* = _____

vii *g* = _____

viii *h* = _____

ix *i* = _____

Soil class description (NZ)	Silt (S)	Sand (Z)	Clay (Cl)
Sand	86.9	*a*	4.8
Loamy sand	69.5	21.5	9.0
Sandy loam	50.4	37.3	*b*
Sandy clay loam	35.7	45.2	19.1
Sandy clay	*c*	35.5	40.6
Silt loam	18.8	53.6	27.6
Loam	50.0	*d*	15.0
Loamy silt	*e*	46.6	11.0
Heavy silt loam	13.1	57.9	*f*
Silty clay loam	*g*	53.6	34.6
Clay loam	18.2	48.2	*h*
Silty clay	10.5	32.4	57.1
Clay	10.1	43.9	46.0
Peat	10.7	*i*	23.0

Table 18.3

b Construct a triangular graph to compare the relationship between the three components of New Zealand's soil.

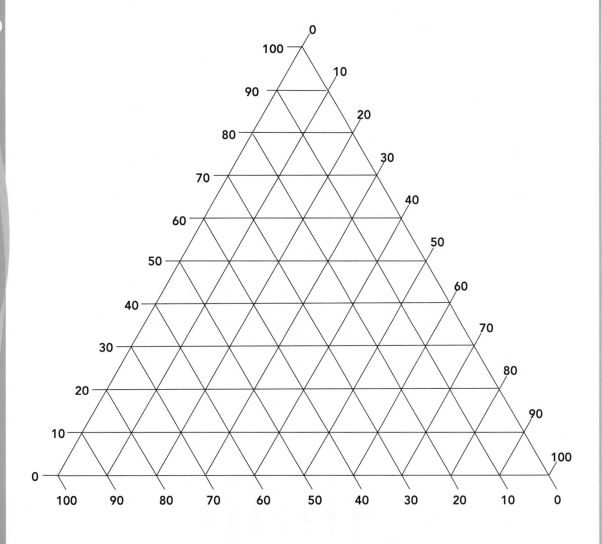

c Write a statement summarising the relationship between the three components of New Zealand's soil.

_____

_____

_____

_____

_____

_____

_____

A pictograph is a pictorial representation of statistical information where each value is represented by a proportional number of pictures or symbols. A pictograph has a form similar to that of a bar graph.

To construct a pictograph, follow the steps below:

1 Decide what information is to be plotted on each axis. In most cases, you will plot the non-quantifiable variable on one axis (e.g. country names, age groups, or periods such as months or years) and quantifiable data on the other.

2 Like bar graphs, pictographs abide by the graphing convention that requires the variable axis to follow a constant number scale starting from zero (e.g. 0, 5, 10, 15 or 0, 10, 20, 30). You will therefore need to determine an appropriate scale for the variable axis.

3 Having determined the range and scale of the data to be plotted, use a ruler to construct the axes. Like most graphs, pictographs have two axes: the x-axis is usually horizontal (i.e. runs across the bottom of the graph) while the y-axis is usually vertical (i.e. runs up the left-hand side of the graph). Ensure each axis is long enough to accommodate the range of data you wish to show.

4 Label each axis (including the units of measurement) and give the graph a title that clearly states what the graph illustrates. If appropriate, the title should also include location and date specific information.

5 Choose a symbol to represent the variable you want to plot. Then decide the quantity that the symbol will represent. For example, in Resource 19.1 the symbol represents 10 million tourists. When plotting the symbols onto your graph, ensure they are drawn with constant spacing and width.

6 If appropriate, label each bar and include a key.

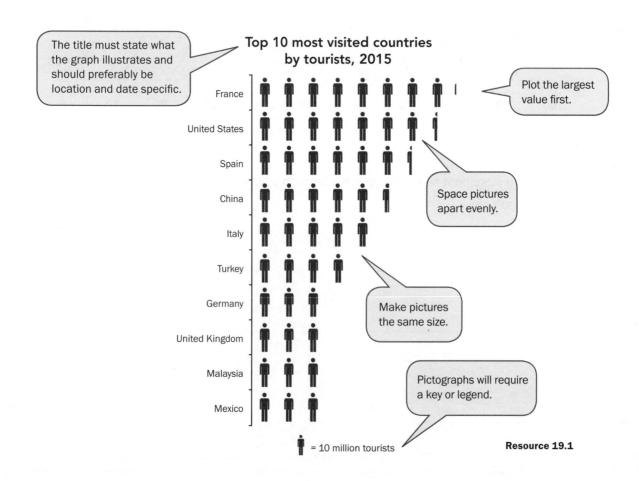

Resource 19.1

1 Study Resource 19.1 and then complete the following activity.

  a Estimate the number of tourists who visited:

  i France _____

  ii China _____

  iii Italy _____

  iv Malaysia _____

2 Using the data in Table 19.1, construct a pictograph to show the number of vehicles per kilometre of road.

Country	Vehicles (per km of road)
South Korea	161
Israel	126
United Kingdom	77
Iran	53
Poland	47

Country	Vehicles (per km of road)
Australia	18
China	13
Rwanda	3
New Zealand	33

Table 19.1

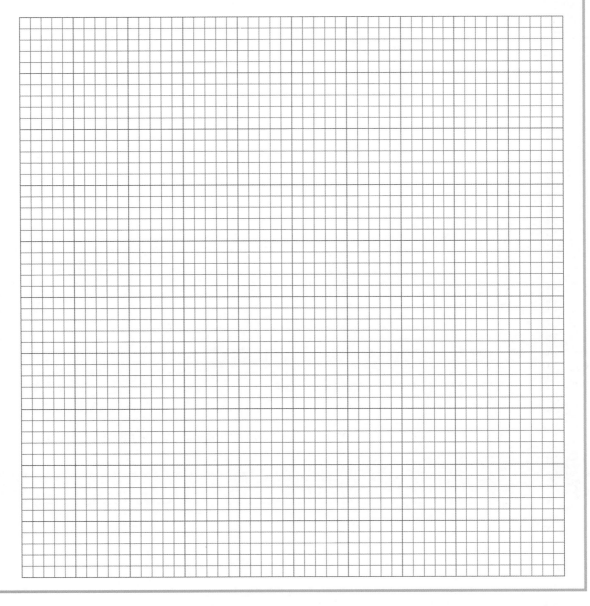

3 Using the data in Table 19.2 to complete the following activities.

a Construct a pictograph to show the total number of military personnel from selected countries in 2015.

Country	Total military personnel (10 000s)	Country	Total military personnel (10 000s)
China	350.3	Iran	91.3
USA	235.0	Mexico	41.7
Russia	336.4	Canada	10.1
Turkey	99.2	Norway	7.2
Iraq	80.3	New Zealand	1.1

Table 19.2

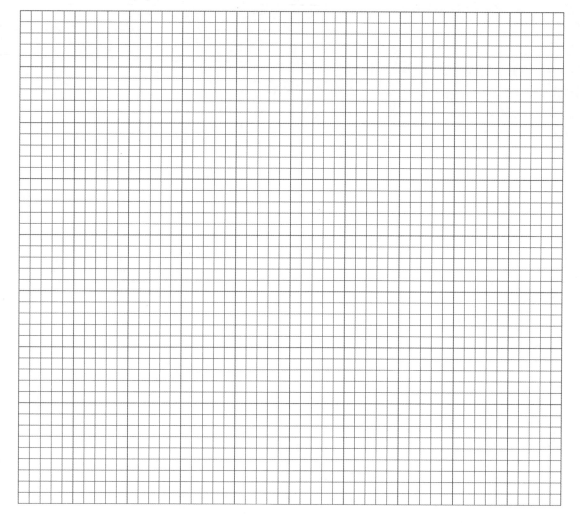

b Write a statement summarising the global distribution of the military personnel.

_____

_____

_____

_____

A climate graph shows average temperature and rainfall experienced at a particular location throughout the year. It consists of a bar graph showing average monthly rainfall and a simple line graph showing average monthly temperature.

A climate graph is constructed using climate data collected over several decades by organisations such as the New Zealand MetService and the National Institute of Water and Atmospheric Research (NIWA). These organisations in turn make the data available to others who rely on accurate climate information to make informed decisions. Viticulturists (wine makers), for example, use climate data to assess frost risk for their crops.

To construct a climate graph, follow the steps below:

1 A climate graph has one horizontal axis banded by two vertical axes. To construct the horizontal axis, divide the axis into 12 even segments to represent the months of the year.

2 Place the rainfall scale (mm) on the left-hand side of the graph and the temperature scale (˚C) on the right-hand side of the graph. Determine an appropriate scale for the two vertical axes and follow a constant number scale starting from zero for each axis (e.g. 0, 5, 10, 15 or 0, 100, 200, 300).

3 Label each axis (including the units of measurement) and give the graph a title that clearly states what the graph illustrates. The title should include location specific information and may even include the location's latitude and longitude.

4 Plot rainfall data for each month and colour each plotted bar blue.

5 Plot the average temperature data for each month with an 'X'. Ensure that each temperature value is positioned at the centre of each month. Join the data points with a red smooth curve.

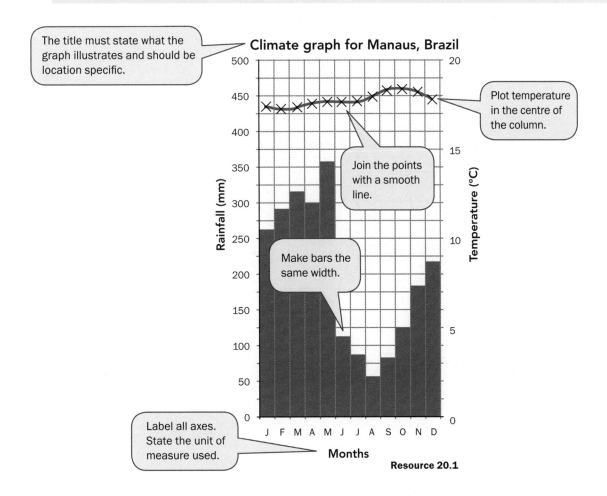

Climate graph for Manaus, Brazil

Resource 20.1

1 Study Resource 20.1 and then complete the following activities.

  a State the temperature and rainfall for:

   i January _____   iii July _____

   ii April _____   iv October _____

  b Identify the wettest and driest months of the year. _____

  c Identify the warmest and coolest months of the year. _____

  d Write a statement summarising Manaus's annual climate.

   _____

   _____

   _____

2 Using the data in Table 20.1, construct a climate graph for San Pedro de Atacama, Chile.

	J	F	M	A	M	J	J	A	S	O	N	D
Rainfall (mm)	1	0	0	4	15	1	0	2	0	0	13	4
Temperature (˚C)	24	24	23	21	19	18	17	17	18	19	20	22

Table 20.1

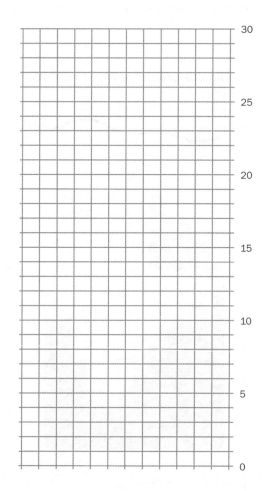

3 What evidence does the climate graph for San Pedro de Atacama contain about the climate of the Atacama Desert?

_____

_____

_____

_____

21 | Population pyramids

A population pyramid is a special type of horizontal bar graph that shows the age-sex structure of a population in one-year, five-year, or ten-year age groups.

The shape of a population pyramid reflects the influence of births, deaths and migration on a population over time and shows whether a population is expanding, stable or likely to decline (Resource 21.1). In general, a population with a high birth rate and low death rate has a broad-based, triangular-shaped pyramid. Populations with low birth rates and low death rates are usually narrower at the base and have straighter sides.

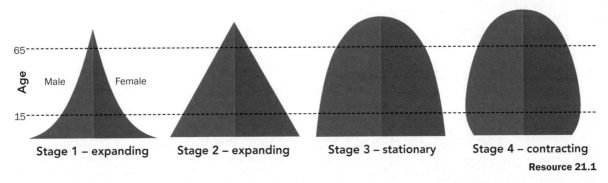

Resource 21.1

To construct a population pyramid, follow the steps below:

1 Use a ruler to construct the axes. Population pyramids have two axes: the horizontal or x-axis usually runs across the bottom of the graph while the vertical or y-axis usually runs up the left-hand side or centre of the graph.

2 Divide the vertical axis into segments to correspond with the age data you are using. Most population pyramids use one-year, five-year, or ten-year age groups.

3 Determine an appropriate scale for the horizontal axis. Note that it is best to construct population pyramids using percentages rather than numbers since this makes it possible to compare countries with different-size populations.

4 Label each axis and give the graph a title that clearly states what the graph illustrates. The title should also include location and date specific information.

5 Beginning at the bottom of the graph, plot the percentage of the population that is 0–4 years and male. Shade this bar on the pyramid and repeat for females, using a different colour. Repeat this step for each age group until the pyramid is complete (Resource 21.2).

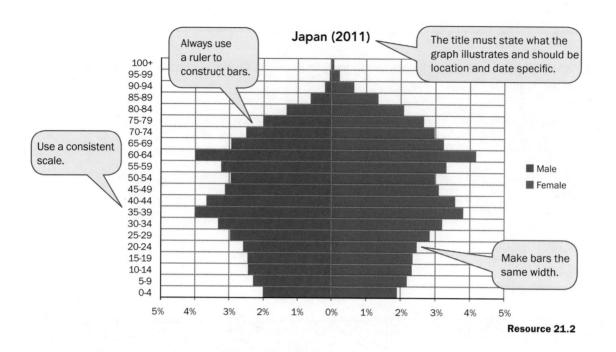

Japan (2011)

Resource 21.2

1 Study Resource 21.2 and then complete the following activities.

a Estimate the percentage of Japanese under the age of 15 years in 2011. _____

b Estimate the percentage of Japanese over the age of 65 years in 2011. _____

c Using Resource 21.1 as a guide, describe the shape of Japan's population pyramid.

_____

_____

_____

2 Use the data in Table 21.1 to construct a population pyramid for Russia.

Age	Male %	Female %
0–4	2.8	2.7
5–9	2.7	2.6
10–14	2.3	2.2
15–19	2.5	2.4
20–24	3.9	3.7
25–29	4.4	4.3
30–34	3.8	3.9
35–39	3.5	3.7
40–44	3.2	3.4
45–49	3.4	3.8
50–54	3.8	4.5
55–59	3.2	4.2
60–64	2.4	3.5
65–69	1.0	1.7
70–74	1.5	3.0
75–79	0.8	1.9
80–84	0.5	1.5
85–89	0.1	0.8
90–94	0.0	0.2
95–99	0.0	0.1
100+	0.0	0.0

Table 21.1

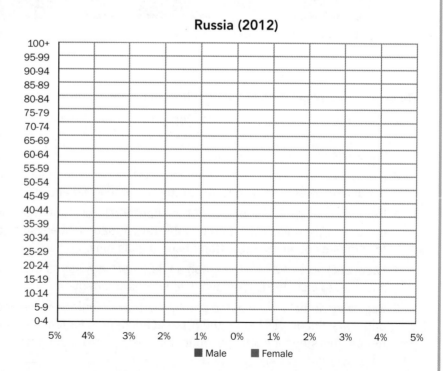

Russia (2012)

Geography Skills for NCEA Level Two
ISBN: 9780170389341

3 What evidence does the population pyramid for Russia contain to suggest that it has an ageing population?

_____

_____

_____

4 Use the data in Table 21.2 to construct a population pyramid for Brazil.

Age	Male %	Female %
0–4	4.4	4.2
5–9	4.4	4.3
10–14	4.4	4.2
15–19	4.2	4.0
20–24	4.1	4.0
25–29	4.2	4.1
30–34	4.2	4.2
35–39	3.8	3.8
40–44	3.5	3.6
45–49	3.1	3.2
50–54	2.6	2.8
55–59	2.1	2.3
60–64	1.6	1.8
65–69	1.2	1.4
70–74	0.8	1.1
75–79	0.5	0.8
80–84	0.3	0.5
85–89	0.1	0.2
90–94	0.0	0.1
95–99	0.0	0.0
100+	0.0	0.0

Table 21.2

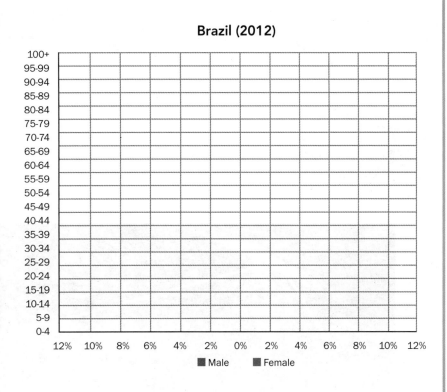

Brazil (2012)

5 What evidence does the population pyramid for Brazil contain to suggest that it has a youthful population?

_____

_____

_____

Visual image interpretation

Visual images can be used to illustrate almost any aspect of geography. As a student of geography, you may on occasion be asked to identify and interpret what you can see in an image, or asked to compare the information from an image with that from a map.

22 | Photograph interpretation

Geographers regularly use photographic images to record information about a setting. Photographic images are an effective tool as they provide a visual record of natural and cultural features of the environment, and help us to understand the way different elements of the environment relate or interact.

Photographs also allow geographers to study how environments change over time. By comparing photographs taken at different times, it is possible to analyse change in any one environment.

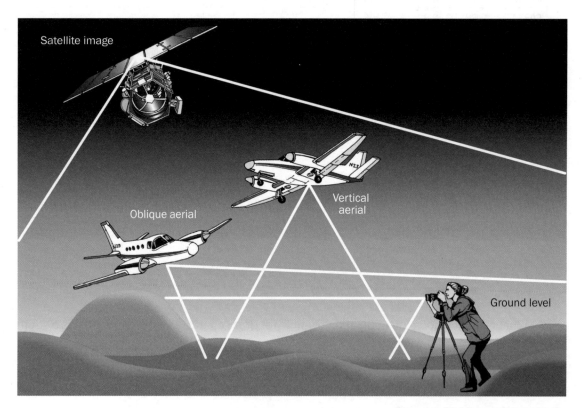

Resource 22.1 Types of photographs

Three types of photographs are useful to geographers:

- **Ground-level photographs** are images taken from the ground to maximise the horizontal view. With ground-level photographs, foreground features appear larger than background features.

- **Aerial photographs** are images taken of the Earth's surface from the air. They show a bird's eye view from either directly above (vertical) or from an oblique angle. Oblique photographs have an advantage over vertical photographs in that they show both the tops and sides of objects, making them easier to identify. The main disadvantage of oblique photographs is that they do not have a consistent scale.

- **Satellite images** are created from data collected from satellites orbiting the Earth. They usually use artificial colours. With satellite images, spatial patterns are clearly visible over a large area (Resource 22.5).

Resource 22.2 Boston, USA

To interpret a photograph, follow the steps below:

1 Determine whether the photograph shows an aerial or ground-level perspective. If the perspective is aerial, determine if its orientation is vertical or oblique.

2 Identify any visual clues that help identify the photograph's setting and location.

3 Identify the main natural (e.g. relief and drainage features, vegetation, soil and climate) and cultural features (e.g. patterns of settlement, transportation networks and other land uses) of the environment shown in the photograph.

4 Examine the way different features shown in the photograph interact. For example, the photograph might show settlement along a coastline or on the flood plain of an adjacent river.

ISBN: 9780170389341

1 Study the photographs in Resources 22.2, 22.3 and 22.4.

Resource 22.3 Farmland near Golden Bay, South Island   **Resource 22.4 Coastal Northland, North Island**

a Identify which photograph is:

i An oblique aerial photograph _____

ii A vertical aerial photograph _____

iii A ground-level photograph _____

2 How do vertical aerial photographs differ from satellite photographs?

_____

_____

_____

_____

_____

_____

3 Study the time series of satellite images of the Dead Sea (Resource 22.5). The image uses artificial colour and shows deep waters as blue or dark blue, while brighter blues indicate shallow waters or salt ponds (in the south). The pale pink and sand-colour regions are barren desert landscapes, while green indicates sparsely vegetated lands. Denser vegetation appears bright red. Near the centre is the Lisan Peninsula, which forms a land bridge through the Dead Sea. For millennia the Dead Sea has been a source of salt, which is culled from the sea and used for water conditioning, road de-icing and the manufacturing of plastics. The growth of expansive salt evaporation projects is clearly visible over the span of 39 years.

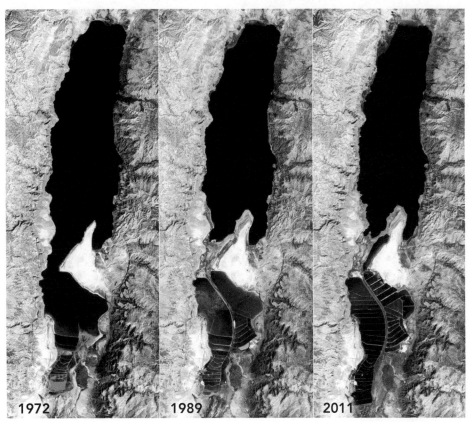

1972   1989   2011

Resource 22.5 The Dead Sea

a Compare the satellite images above and describe the main land use changes that have taken place in and around the Dead Sea between 1972 and 2011.

_____

_____

_____

_____

_____

_____

_____

_____

b What process has been the driving force behind the change? Justify your answer.

_____

_____

_____

_____

_____

_____

_____

_____

ISBN: 9780170389341

23 | Précis sketches

Geographers regularly use sketches to identify and record features of the landscape. If a sketch is drawn from a photograph, then it is called a précis sketch (which simply means a summary sketch).

You do not need artistic ability to draw an effective précis sketch. You do, however, need to be able to illustrate your understanding of the landscape shown in the photograph, by being able to identify and draw the important geographic features found within it.

To construct a précis sketch from a photograph, follow the steps below:

1 Study the photograph and choose the area to be included in the sketch.

2 Draw a frame in the same shape as the photograph you wish to sketch. If you are drawing a précis sketch from an oblique aerial photograph, then it may be beneficial to construct a trapezoid-shaped frame with a narrow base, to compensate for the narrow field of vision in the foreground.

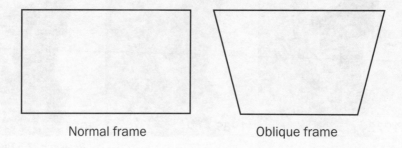

Normal frame   Oblique frame

3 Divide both the photograph and the précis sketch frame into three equal areas: foreground, middle distance and background (Resource 23.1).

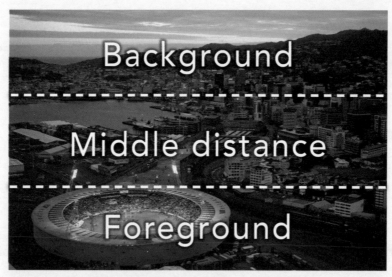

Resource 23.1 Dividing the frame into equal thirds will help you with your sketch

4 Select the main natural and cultural features shown in each area of the photograph and sketch the outline of their shape into the frame of the précis sketch.

5 Use colour shading to highlight main geographical features.

6 Label the main geographical features.

Resource 23.2 Marlborough, South Island

Resource 23.3 Sample précis sketch of Marlborough

ISBN: 9780170389341

1 Explain why geographers construct précis sketches from photographs.

_____

_____

2 Construct a précis sketch for Resource 23.4. Label the main geographical features illustrated.

Resource 23.4

3 Construct a précis sketch for Resource 23.5. Label the main geographical features illustrated.

Resource 23.5

ISBN: 9780170389341

Geographers have long acknowledged the value of cartoons as a way of fostering an understanding of contemporary geographic issues.

The interpretation of cartoons requires an appreciation of the various techniques a cartoonist uses to convey or communicate an idea. For example, the cartoon by Guy Keverne Body entitled 'Whanganui', demonstrates how several techniques can be combined into one cartoon to communicate an idea of geographical importance.

When interpreting cartoons in a geographical context, three questions should be asked of the cartoon:

1 What information does the cartoon convey? This includes identifying the contemporary geographic issue portrayed in the image, and the perspective or viewpoint of the cartoonist.

2 What geographical concepts or ideas are addressed in the cartoon?

3 What are the geographical implications of the issue addressed by the cartoonist?

Use of exaggeration and distortion: Note the exaggeration of facial features to identify the ethnicity of the person.

Use of humour.

Visual metaphor: The broken bridge signifies the breakdown in Maori-Pakeha relations.

Use of caricature: The dress and skin colour of the cartoon character clearly identifies the individual's ethnic group as Maori.

Use of symbolism: A bridge crossing shaped in the letter H symbolises the potential rift in Maori-Pakeha relations.

Resource 24.1

Context: The cartoon shows the former mayor of Wanganui, Michael Laws, on one side of the Whanganui River driving an excavator with which he is removing the centre portion of the 'H' that forms a bridge across the river. On the other side of the river is co-leader of the Maori Party, Tariana Turia, looking very frustrated. The cartoon refers to the debate over putting the 'h' into 'Wanganui'.

1 Study Resource 24.2 and complete the following activities.

CHCH CBD Development...

INNOVATION · PROGRESS · AFFORDABILITY · BUSINESS DEVELOPMT · CATHEDRAL PLANS · CONFIDENCE · CREATIVITY

AL NISBET

Resource 24.2

Context: Refers to delays in major projects for redevelopment of Christchurch central business district.

a Identify the issue addressed in the cartoon.

_____

_____

b Describe the geographical processes relevant to the issue.

_____

_____

c Outline at least two different perspectives relevant to the issue addressed in the cartoon.

_____

_____

_____

_____

ISBN: 9780170389341

d Identify the actions that individuals, groups and governments can take to address the issue highlighted by the cartoonist.

_____

_____

_____

_____

_____

2 Study Resource 24.3 and complete the following activities.

Resource 24.3

Context: Depicts two men standing over a dead Hector's dolphin. Refers to the threat posed to the endangered species by set nets.

a Identify the issue addressed in the cartoon.

_____

_____

_____

b Describe the geographical processes relevant to the issue.

_____

_____

_____

c Outline at least two different perspectives relevant to the issue addressed in the cartoon.

_____

_____

_____

_____

_____

d Identify the actions that individuals, groups and governments can take to address the issue highlighted by the cartoonist.

_____

_____

_____

_____

3 Using the Internet, locate two cartoons that address a geographical issue such as global inequalities, globalisation, population growth, pollution or land degradation.

a Analyse each cartoon using the steps outlined above.

b Share your findings with the rest of the class and mount a wall display of the cartoons and your analysis.

Geographic models

A model is a simplified representation of something that is real. You have most likely seen and used models in the past. A globe, for instance, is a model of the Earth. Geographers use models to analyse geographic processes when the real object of study is too large to examine, or where the processes that created it operate over too long a period.

Geographical models may be simple conceptual models such as a box and arrow diagram showing the flows of energy between components of a system, or complex computer-based mathematical models. Physical geographers, for example, construct physical models like stream tables to investigate the impact of river or fluvial processes on Earth's surface. Climate scientists, on the other hand, use complex mathematical models to predict changes in climate over time.

In this chapter you will learn about four conceptual models that are important to the study of Level 2 NCEA Geography.

25 | Models of urban structure

Throughout the twentieth century numerous sociologists, economists and geographers tried to create models to explain variations in the form of cities. While every city has its own unique form, studies have shown that most urban areas share similar characteristics. As a result, several models explaining urban structure have been put forward. The following diagrams describe the important characteristics, assumptions and limitations of three of the most popular models of urban structure.

Burgess's Concentric Model

The Concentric Model was the first to identify the distribution of socio-economic groups within urban areas. Based on the US city of Chicago, it was created by sociologist E. Burgess in 1924. According to the model, a city grows outward from a central location in a series of concentric zones. The inner-most zone represents the central business district (CBD). It is surrounded by a zone of transition or twilight zone, which contains a mix of light industry and older housing largely inhabited by the city's poorest inhabitants. The third concentric zone contains low-class housing and is usually occupied by those who have migrated from the zone of transition. The fourth zone has higher quality houses usually occupied by the middle class. The outermost zone is called the commuter's zone. This zone is inhabited by the wealthy, who can afford to live in high-quality residential suburbs and take the daily commute into the CBD to work.

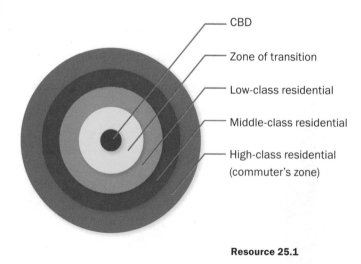

- CBD
- Zone of transition
- Low-class residential
- Middle-class residential
- High-class residential (commuter's zone)

Resource 25.1

Hoyt's Sector Model

In 1939, economist H. Hoyt suggested a second model of urban structure. His model, coined the Sector Model, proposed that a city develops in sectors instead of zones. Recognising that different areas of a city attract different activities, as a city grows and these activities flourish and expand outward, they do so in a wedge that becomes a sector of the city.

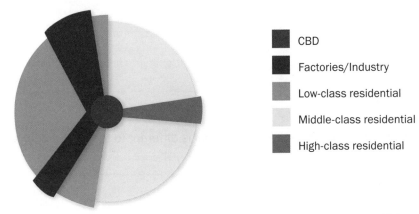

CBD

Factories/Industry

Low-class residential

Middle-class residential

High-class residential

Resource 25.2

Harris and Ullman's Multiple Nuclei Model

Geographers C.D. Harris and E.L. Ullman developed the Multiple Nuclei Model in 1945. According to this model, a city is said to grow outwards from more than one central location or node. The model suggests that certain activities are attracted to particular nodes, while others try to avoid them. For example, a governmental node may attract embassies, well-educated residents, administrative services and quality restaurants, whereas a seaport may attract transport services and warehouses.

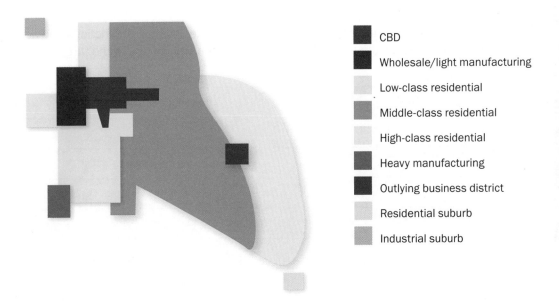

CBD

Wholesale/light manufacturing

Low-class residential

Middle-class residential

High-class residential

Heavy manufacturing

Outlying business district

Residential suburb

Industrial suburb

Resource 25.3

ISBN: 9780170389341

1   Describe the main features of the Burgess model and describe why it is useful to geographers.

_____

_____

_____

_____

_____

_____

2   Explain how the Harris and Ullman model is different to the Hoyt model of urban structure.

_____

_____

_____

_____

_____

_____

_____

_____

_____

3   With reference to a named city, discuss the structure (i.e. location of functions and activities) of the city and discuss the extent to which your chosen city fits one of the three models.

_____

_____

_____

_____

_____

_____

_____

_____

_____

_____

_____

_____

_____

The core-frame model is designed to show the urban structure (land use and functions) of the central business district (CBD) of a town or city.

The model includes an inner core (pink) where land is expensive and used intensively, resulting in the high-rise development. This area is the focus of transport systems and has a concentrated daytime population of workers. The outer core (yellow) and frame (blue) have lower land values and are less intensively developed than the inner core. The zone of assimilation (expansion and growth) and zone of discard (retreat and decline) are together called the zone of transition and reflect the areas of the CBD that are undergoing change (Resource 26.1).

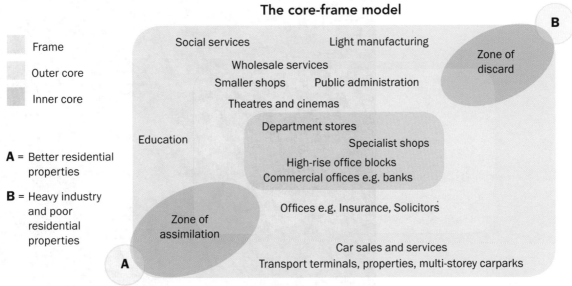

The core-frame model

Frame

Outer core

Inner core

A = Better residential properties

B = Heavy industry and poor residential properties

Social services · Light manufacturing
Wholesale services
Smaller shops · Public administration
Theatres and cinemas
Department stores
Specialist shops
High-rise office blocks
Commercial offices e.g. banks
Education
Offices e.g. Insurance, Solicitors
Zone of assimilation
Zone of discard
Car sales and services
Transport terminals, properties, multi-storey carparks

A

B

Resource 26.1

The core-frame model allows geographers to identify and map the seven main characteristics of the CBD:

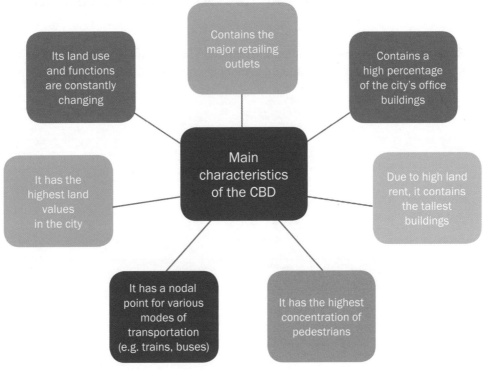

Main characteristics of the CBD

- Its land use and functions are constantly changing
- Contains the major retailing outlets
- Contains a high percentage of the city's office buildings
- It has the highest land values in the city
- Due to high land rent, it contains the tallest buildings
- It has a nodal point for various modes of transportation (e.g. trains, buses)
- It has the highest concentration of pedestrians

Resource 26.2

1 Study the aerial photograph of Auckland's CBD. Locate and shade the:

   a Inner core

   b Outer core

   c Frame

   d Zone of transition

   e Zone of assimilation

   f Zone of discard

2 Rank each of the characteristics in Resource 26.1 in order of importance when it comes to identifying the functional land use zones of your town or city's CBD. Justify your decision.

_____

_____

_____

_____

_____

_____

_____

3 Use the Internet (e.g. Google Earth) to source a land use map of your town or city. Draw a précis map of your CBD in the space below and locate and shade the:

a   Inner core

b   Outer core

c   Frame

d   Zone of transition

e   Zone of assimilation

f   Zone of discard

Key:

☐ Inner core

☐ Outer core

☐ Frame

☐ Zone of transition

☐ Zone of assimilation

☐ Zone of discard

ISBN: 9780170389341

One of the earliest models to account for economic growth was developed by economist and political theorist W.W. Rostow in 1960. Following a study of 15 mainly European countries, Rostow proposed that all countries had the potential to break the cycle of poverty and develop through five stages of development (Resource 27.1):

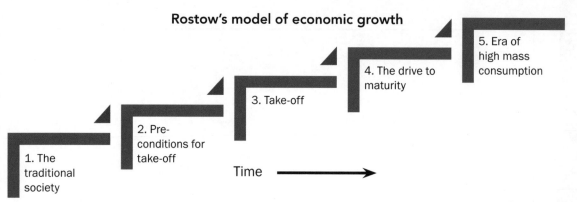

Rostow's model of economic growth

5. Era of high mass consumption

4. The drive to maturity

3. Take-off

2. Pre-conditions for take-off

1. The traditional society

Time

Resource 27.1

However, Rostow's model is not without its critics and is regarded by some geographers as outdated and oversimplified. Other criticisms include:

- The model assumes all countries begin at a Stage 1 level of development.

- Some countries, especially in Africa, have struggled to take-off despite injections of money (a precondition for take-off) in the form of aid due to mounting debt repayments.

- The model ignores the effects of colonialism and imperialism, where ruling countries are often developed at the expense of others.

- It ignores the likelihood that less-developed countries have the potential to learn and gain from the experiences and technologies of other more-developed countries. In other words, the model ignores the effects of globalisation.

- The model cannot be universally applied.

In response to these criticisms, geographers Barke and O'Hare in their 1992 thesis put forward an alternative model of economic development, for countries that had yet to industrialise. Barke and O'Hare's model differed to that of Rostow in that it included four stages of industrial growth where the characteristics of each stage can potentially exist side by side, in effect creating a 'dual' economy (Resource 27.2).

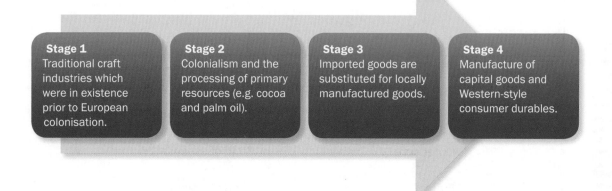

Stage 1
Traditional craft industries which were in existence prior to European colonisation.

Stage 2
Colonialism and the processing of primary resources (e.g. cocoa and palm oil).

Stage 3
Imported goods are substituted for locally manufactured goods.

Stage 4
Manufacture of capital goods and Western-style consumer durables.

Resource 27.2 Barke and O'Hare's model for West Africa

1 Study Resource 27.1 and draw a line to match each stage of development with the correct description.

Stage 1	Economic growth becomes self-sustaining. Industry begins to increase and diversify. More complex transport systems develop.
Stage 2	Subsistence economy mainly based on farming, with limited technology or capital.
Stage 3	Rapid expansion of tertiary industries and welfare facilities. Employment declines in the manufacturing industry but increases in the service sector.
Stage 4	Manufacturing industries grow rapidly. Airports, seaports, roads and railways are built to facilitate trade. Agriculture becomes commercialised and more mechanised.
Stage 5	Country receives an injection of foreign capital (cash). Mining and forestry industries develop.

2 Study the following table of employment structure based on Rostow's model. Use the Internet (e.g. data.un.org or data.worldbank.org) to identify countries representative of each of the five stages.

	Primary sector (extraction)	Secondary sector (manufacturing)	Tertiary sector (services)
Stage 1	large majority	very few	very few
Stage 2	large majority	few	very few
Stage 3	declining	rapid growth	few
Stage 4	few	stable	rapid growth
Stage 5	very few	declining	large majority

a Stage 1: _____

b Stage 2: _____

c Stage 3: _____

d Stage 4: _____

e Stage 5: _____

3 With reference to Rostow's model, describe the changes a country will go through as it progresses through each stage.

_____

_____

_____

_____

_____

_____

_____

_____

_____

_____

_____

_____

4 Barke and O'Hare's model of development for West Africa shows that some sectors of poorer countries pass through four stages of industrial growth. Briefly describe the characteristics of each stage.

_____

_____

_____

_____

_____

_____

_____

_____

_____

_____

_____

_____

_____

The core-periphery model (Resource 28.1) was developed in 1966 by John Friedmann to explain spatial variations in growth and economic development. In the model, the core forms the most prosperous and developed part of a country or region. As economic activity and development usually decrease with distance from the core, the periphery is relatively poorer and less developed. In between the core and periphery, the semi-periphery is the part that is currently experiencing the most change as it strives for core status. The core-periphery model works on many scales, from towns and cities to a global scale.

Friedmann's core-periphery model

Core: capital city, chief port, major industries and urban areas, most services and investment.

Semi-periphery: includes NICs and cities that have moved from being part of the periphery.

Periphery: levels of wealth, development and standards of living decrease with distance from the core; fewer jobs and services, less investment.

Resource 28.1

The model operates on the premise that as a country develops, two processes are likely to occur:

1 Economic activity at the core will grow at a faster rate than the periphery due to its ability to attract new industries and services (e.g. finance, communication and insurance companies). As a result, the core will be able to afford facilities such as hospitals, schools, housing and shopping centres which in turn will generate employment opportunities. These facilities and services act as pull factors and encourage inward migration from the periphery.

2 After a period of time, industry and wealth will spread to secondary cores. This can sometimes result in the decline of the original core.

When applied to the global setting, the majority of economic activity is concentrated in a core region of wealthy countries (e.g. Western Europe, USA, Canada, Japan, Australia and New Zealand). In contrast, the global periphery (e.g. Africa, South America, most of Asia) is characterised by lower living standards and extreme poverty. The disparity in wealth between the global core and the global periphery is significant in that the core, with no more than 15% of the world's population, generates more than 75% of the world's annual income.

ISBN: 9780170389341

1 Study Resource 28.2 of the European Union below, which shows household income per person.

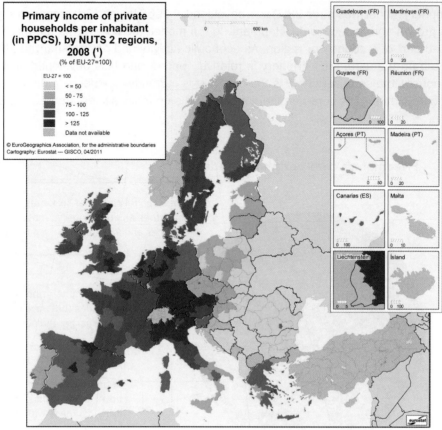

Primary income of private households per inhabitant (in PPCS), by NUTS 2 regions, 2008 (¹)
(% of EU-27=100)

EU-27 = 100
- <= 50
- 50 - 75
- 75 - 100
- 100 - 125
- > 125
- Data not available

© EuroGeographics Association, for the administrative boundaries
Cartography: Eurostat — GISCO, 04/2011

Resource 28.2

a Identify the countries of Europe that could be considered part of the:

i core _____

ii periphery _____

iii semi-periphery _____

b Describe the distribution of economic development in the European Union.

_____

_____

_____

2 With reference to the core-periphery model, explain why populations often migrate from the periphery to the core.

_____

_____

_____

_____

_____

Numeracy in geography

Numeracy refers to the skill of being able to understand and work with numbers. However, just because numeracy deals with numbers does not mean that its use is limited to the subject of mathematics. Geographers routinely use numbers to help them identify and interpret patterns and trends in geographic phenomena.

For example, geographers use numbers to calculate:

- Population growth, size and density in a certain place or area. Knowing the population characteristics of a place helps governments and non-government organisations (NGOs) plan for the future.

- The scale of a map relative to the real world.

- The coordinates of latitude and longitude and grid references on maps to find a specific location in the world.

- Temperature and rainfall in a particular location to help farmers predict future rainfalls. This is important so that farmers and market gardeners do not waste water through poorly planned irrigation.

The ability to interpret and then analyse statistical information is an essential skill for a geographer to possess. In Level 2 NCEA Geography, you need to know how to work with numbers to calculate means, modes, ranges and percentage change.

29 | Working with numbers: calculating mean, mode and range

The most fundamental of numeracy skills is the ability to understand and calculate statistical means, modes and ranges.

Mean

When working with a data set, the mean (or arithmetic average) is calculated by finding the sum of all the given values divided by the total number of values. It is expressed by formula as:

$$\bar{x} = \frac{\sum x}{n}$$

Where $\bar{x}$ = mean, $\sum$ = the sum of, x = the value of an individual variable and n = the number of values in the data set. Put in another way, the formula can be presented as a statement:

mean = sum of all the observed values ÷ number of observations

For example, in a village of 10 households the number of people living in each household is 7, 5, 0, 7, 8, 5, 5, 4, 5 and 2.

Mean   = (7 + 5 + 0 + 7 + 8 + 5 + 5 + 4 + 5 + 2) ÷ 10
         = 48 ÷ 10
         = 4.8

Mode

The mode is the most frequently observed data value. There may be no mode if no value appears more than any other.

mode = the most frequently observed data value

In the previous example, the value of the mode is 5, as four households contain five inhabitants. This number occurs more frequently than any other data value.

Range

The data range refers to the difference between the highest and the lowest observed data value.

range = the highest observed data value − the lowest observed data value

In our previous example, the range would be calculated as follows:

Range = 8 − 0

= 8

1 Study the data in the table below, showing average monthly temperature readings (°C) for Auckland and Dunedin. For each city calculate the annual:

Location	J	F	M	A	M	J	J	A	S	O	N	D
Auckland	19	20	19	16	14	12	10	11	13	14	16	18
Dunedin	15	15	14	12	9	7	7	8	9	11	12	14

	Auckland	Dunedin
a Mean		
b Mode		
c Range		

2 Study the data in the table below, showing life expectancy (years) in Central Africa, and calculate the:

Region	Life expectancy at birth (2016)
Angola	52
Cameroon	55
Central African Republic	51
Chad	52
Congo, Dem. Rep.	57
Congo, Rep.	62
Equatorial Guinea	58
Gabon	64
Sao Tome and Principe	66

a Mean

b Mode

c Range

Working with numbers: calculating percentage change

When working with statistics, it is sometimes useful to be able to quantify the rate of proportional change between an original value and its new value. A popular mathematical technique for measuring change between two values is to calculate its percentage change.

To calculate percentage, apply the following formula:

$$\text{percentage change} = \frac{\text{(difference between the two values)}}{\text{original value}} \times \frac{100}{1}$$

For example, in 1960 New Zealand's population was 2.37 million. In 2010, it was estimated at 4.38 million. The percentage increase between 1960 and 2010 would be calculated as follows:

$$\text{percentage change} = \frac{\text{(4.38 million} - \text{2.37 million)}}{\text{2.37 million}} \times \frac{100}{1}$$

$$= 84\%$$

Learning Activities

1956		2013	
Population	2 174 061	Population	4 242 048
Median age	29	Median age	38
Median income (pounds)	£715	Median income (dollars)	$28 500
Born in NZ	85.8%	Born in NZ	74.8%
Living in Auckland	24.2%	Living in Auckland	34%
Living in North Island	67%	Living in North Island	77%
Ethnicity		**Ethnicity**	
European	92.7%	European	74.0%
Maori	6.3%	Maori	14.9%
Pacific Island	0.4%	Pacific Island	7.4%
		Asian	11.8%

Table 30.1

1 Study Table 30.1, showing the composition of New Zealand's population in 1956 and 2013. Calculate the percentage change between 1956 and 2013 for each population characteristic.

a Population change

_____

_____

b Change in the median age

_____

_____

c Percentage change born in New Zealand

_____

_____

_____

_____

d Percentage change living in Auckland

_____

_____

_____

_____

e Percentage change living in the North Island

_____

_____

_____

_____

f Change in European ethnicity

_____

_____

_____

_____

g Change in Maori ethnicity

_____

_____

_____

_____

h Change in Pacific Island ethnicity

_____

_____

_____

_____

Putting it all together

The learning activities that follow will provide you with an opportunity to practise the skills you have learned throughout this book within the context of an important NCEA Level 2 topic: Demonstrate geographic understanding of differences in development.

The seventeen Sustainable Development Goals (SDGs), which range from ending poverty to achieving gender equality and empowerment by the target date of 2030, are a set of international development goals that have been adopted as a framework for global development by the member countries of the United Nations (UN). As a member of the UN and a signatory to the SDGs, New Zealand is therefore committed to achieving the seventeen goals not only within New Zealand but also abroad.

The seventeen global goals are:

The questions that follow provide an opportunity for you to investigate a selection of the SDGs, their progress to date, and their contribution to global development. By completing these questions, you will discover that our progress thus far towards achieving each goal varies from goal to goal. While we have made significant progress in some areas, significant disparities remain. Therefore, your final task (page 152) will be to summarise the suitability of the SDGs as a framework for making the world a better place.

ISBN: 9780170389341

Development Goal: Zero hunger

The specific aim of Goal 2 is to end hunger, achieve food security and improved nutrition and promote sustainable agriculture by 2030.

1 The data that follows indicates the progress so far, in meeting this goal.

Country	Undernourishment as a percentage of the population (2015)	Country	Undernourishment as a percentage of the population (2015)
Algeria	5	Liberia	32
Angola	14	Madagascar	33
Benin	8	Malawi	21
Botswana	24	Mali	5
Burkina Faso	21	Mauritania	6
Cameroon	10	Mauritius	5
Central African Republic	48	Morocco	5
Chad	34	Mozambique	25
Congo, Rep.	31	Namibia	42
Cote d'Ivoire	13	Niger	10
Djibouti	16	Nigeria	7
Ethiopia	32	Rwanda	32
Gabon	5	Senegal	10
Gambia, The	5	Sierra Leone	22
Ghana	5	South Africa	5
Guinea	16	Tanzania	32
Guinea-Bissau	21	Togo	11
Kenya	21	Uganda	26
Lesotho	11	Zambia	48
		Zimbabwe	33

Resource 31.1

a Using the data in Resource 31.1, construct a choropleth map to show the percentage of people in sub-Saharan Africa who are undernourished.

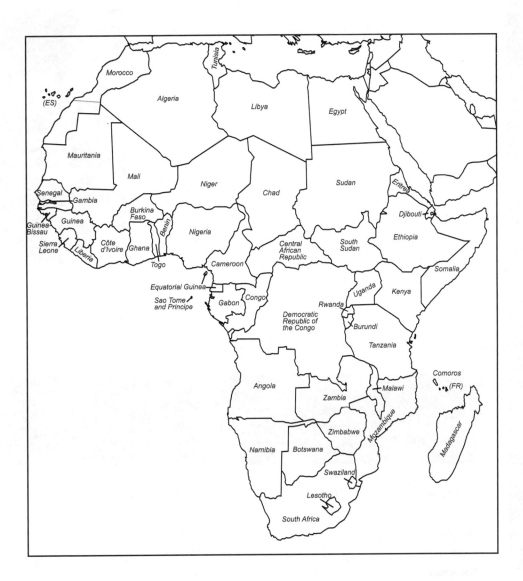

Key:

☐ ☐

☐ ☐

☐ ☐

b With reference to named countries, describe the pattern in the map.

_____

_____

_____

_____

ISBN: 9780170389341

c What evidence does Resource 31.2 contain to suggest the child in the photograph is living in poverty?

_____

_____

_____

_____

_____

_____

_____

_____

Resource 31.2

2 The table below shows the population of sub-Saharan Africa living on less than a US$1.90 per day relative to each country's GDP per capita (US$).

Region	GDP based per capita (2014) (US$)	Percentage of population living on less than US$1.90 per day (2015)	Region	GDP based per capita (2014) (US$)	Percentage of population living on less than US$1.90 per day (2015)
Benin	903	53	Guinea-Bissau	568	67
Botswana	7123	18	Lesotho	1034	60
Burkina Faso	713	55	Liberia	458	69
Burundi	286	78	Madagascar	449	82
Cabo Verde	3641	18	Malawi	255	71
Cameroon	1407	29	Mali	705	49
Central African Republic	359	66	Mozambique	586	69
			Namibia	5408	23
Chad	1025	38	Niger	427	50
Comoros	810	14	Nigeria	3203	54
Congo, Dem. Rep.	442	77	Rwanda	696	60
			Senegal	1067	38
Congo, Rep.	3147	29	South Africa	6484	17
Cote d'Ivoire	1546	29	Swaziland	3477	42
Ethiopia	574	34	Tanzania	955	47
Gabon	10772	8	Togo	635	54
Gambia, The	441	45	Uganda	715	33
Ghana	1442	25	Zambia	1722	64
Guinea	540	35			

Resource 31.3

a Use the data in the table to construct a scatter graph to show the relationship between earnings (GDP) per person (US$) and the percentage of those in each country living in poverty (less than US $1.90 per day).

b Account for the correlation shown in the graph.

_____

_____

_____

_____

3 Describe the overall pattern of malnutrition in sub-Saharan Africa. Explain how the indicators in the questions above are linked.

_____

_____

_____

_____

_____

_____

_____

_____

_____

_____

_____

_____

Development Goal: Quality education

The aim of Goal 4 is to ensure inclusive and equitable quality education and promote lifelong learning opportunities for all.

1 Study the data below, and complete the following activities.

Region	Children out of primary school	
	2003	2013
East Asia & Pacific	8 569 020	6 567 060
Europe & Central Asia	1 370 718	1 349 685
Latin America & Caribbean	2 853 748	4 065 566
Middle East & North Africa	3 851 090	2 183 238
North America	1 420 729	1 909 338
South Asia	23 869 308	10 258 789
Sub-Saharan Africa	39 032 016	32 921 700

Resource 31.4

a Construct two pie graphs to show the change in distribution of out-of-school children from 2003 to 2013.

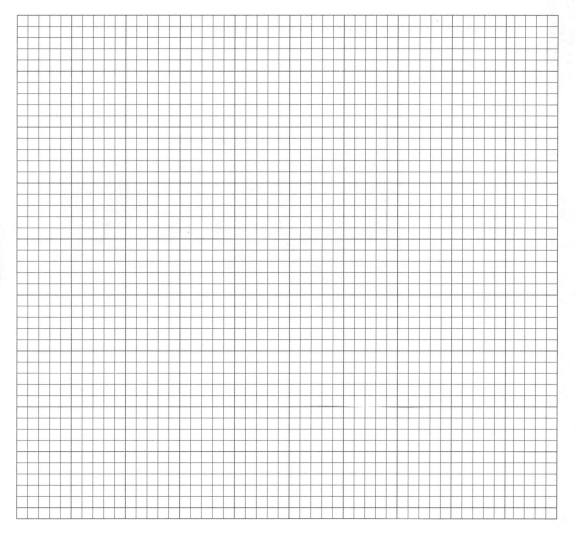

b Describe the change in distribution of children out-of-school between 2003 and 2013.

_____

_____

_____

_____

c Outline the progress that has been made towards the global goal of achieving inclusive education.

_____

_____

_____

_____

_____

Development Goal: Gender equality

Goal 5 seeks to promote gender equality and empower all women and girls.

1 With this goal in mind, Resource 31.5 shows the changing proportion of men and women who have sat on the European Parliament from its inception in 1979 up to the current term; reflecting the the European Union's efforts to achieve this goal.

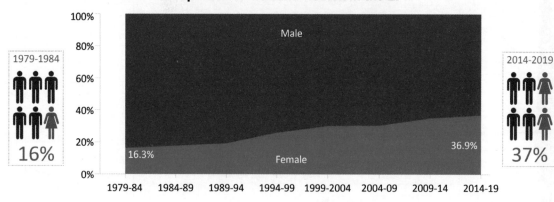

Proportion of men and women in the EP

Resource 31.5

a With reference to Resource 31.5, describe the change in the proportion of men and women in the European Parliament.

_____

_____

_____

_____

b Explain how more women in Parliament supports the global goal of achieving gender equality and empowerment.

_____

_____

_____

_____

_____

_____

_____

_____

_____

Development Goal: Good health and well-being

Goal 3 aims to ensure healthy lives and promote well-being for all regardless of age. In particular, the health of women during childbirth and child health remains a major concern.

1 Study Resource 31.6 and then complete the following activities.

Region	Maternal mortality per 100 000 live births				
	1995	2000	2005	2010	2015
East Asia & Pacific	129	113	94	74	59
Europe & Central Asia	43	33	27	19	16
Latin America & Caribbean	117	99	88	81	67
Middle East & North Africa	138	113	100	89	81
North America	11	12	13	14	13
South Asia	476	388	296	228	182
Sub-Saharan Africa	928	846	717	625	547

Resource 31.6

a Construct a multiple-line graph to show the change in maternal mortality (per 100 000 live births) from 1995 to 2015.

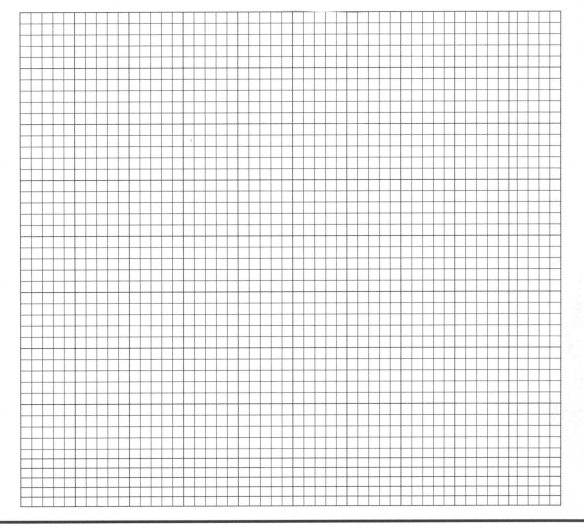

b Describe the change in maternal mortality by region from 1995 to 2015.

_____

_____

_____

_____

Under-five mortality rate (probability of dying by age five per 1000 live births), 2015

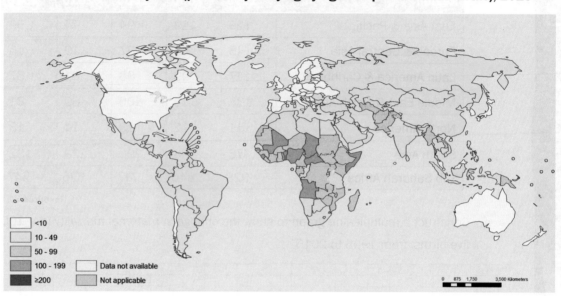

<10
10 - 49
50 - 99
100 - 199   Data not available
≥200   Not applicable

0   875 1,750   3,500 Kilometers

Resource 31.7

2 Study Resource 31.7 and then complete the following activities.

a Identify the regions with the highest under-five mortality rate.

_____

_____

b Identify the regions with the lowest under-five mortality rate.

_____

_____

c With reference to named regions, describe the global differences in under-five mortality.

_____

_____

_____

_____

_____

_____

_____

Geography Skills for NCEA Level Two
ISBN: 9780170389341

Development Goal: Sustainable cities and communities

The aim of Goal 11 is to make cities and human settlements inclusive, safe, resilient and sustainable. Related to this is the specific target of improving the lives of at least 100 million slum dwellers by 2030.

1 Construct a précis sketch for the oblique photograph of Favela da Rocinha, the biggest slum in Rio de Janeiro, Brazil. Label the main geographical features illustrated.

2 Using evidence from the photo, suggest reasons why living conditions in Favela da Rocinha may not be inclusive, safe, resilient and sustainable.

_____

_____

_____

_____

_____

_____

_____

_____

Development Goal: Industry, Innovation and Infrastructure

Of its many aims, one important aspect of Goal 9 is to make available the benefits of new technologies, especially information and communications, to poorer countries.

1 The map below shows the current interconnectedness of the world's telecommunication and Internet networks.

The Internet's undersea world

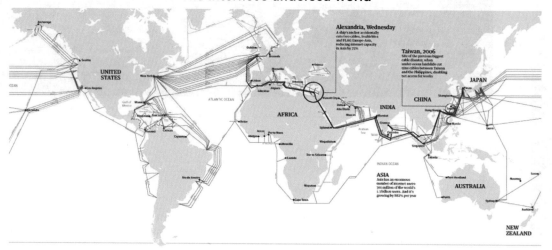

Resource 31.8

a With reference to Resource 31.8, identify which regions are well connected and which are not. Give reasons for the pattern of interconnectedness.

_____

_____

_____

_____

_____

_____

b How does a region's interconnectedness relate to its ability to develop?

Development Goal: Responsible consumption and production

Sustainable Development Goal 12 aims to ensure sustainable consumption and production.

1 Study Resource 31.9 and complete the following activities.

Resource 31.9

a What contemporary geographic issue does the cartoon portray?

_____

_____

_____

_____

_____

_____

b What are the geographical implications of the issue addressed by the cartoonist?

_____

_____

_____

_____

_____

_____

_____

_____

Applying geographic concepts: Interaction

Interaction involves elements of an environment affecting each other and being linked together. Interaction incorporates movement, flow, connections, links and interrelationships. Landscapes are the visible outcome of interactions. Interaction can bring about environmental change.

FROM ASPIRATION TO ACHIEVEMENT:

Breaking down the UN Sustainable Development Goals

PYXERA Global

GOAL 10
Reduce inequality within and among countries

GOAL 16
Promote peaceful and inclusive societies for sustainable development, provide access to justice for all and build effective, accountable and inclusive institutions at all levels

GOAL 5
Achieve gender equality and empower all women and girls

GOAL 4
Ensure inclusive and equitable quality education and promote lifelong learning opportunities for all

GOAL 1
End poverty in all its forms everywhere

GOAL 8
Promote sustained, inclusive and sustainable economic growth, full and productive employment and decent work for all

GOAL 7
Ensure access to affordable, reliable, sustainable and modern energy for all

GOAL 9
Build resilient infrastructure, promote inclusive and sustainable industrialization and foster innovation

GOAL 12
Ensure sustainable consumption and production patterns

GOAL 14
Conserve and sustainably use the oceans, seas and marine resources for sustainable development

GOAL 6
Ensure availability and sustainable management of water and sanitation for all

GOAL 2
End hunger, achieve food security and improved nutrition and promote sustainable agriculture

GOAL 3
Ensure healthy lives and promote well-being for all at all ages

GOAL 11
Make cities and human settlements inclusive, safe, resilient and sustainable

GOAL 13
Take urgent action to combat climate change and its impacts

GOAL 15
Protect, restore and promote sustainable use of terrestrial ecosystems, sustainably manage forests, combat desertication, and halt and reverse land degradation and halt biodiversity loss

HUMAN RIGHTS

ECONOMIC OPPORTUNITY/ EMPLOYMENT

GLOBAL PARTNERSHIPS*

HEALTH

HUMAN & NATURAL ENVIRONMENT

* GOAL 17
Strengthen the means of implementation and revitalize the global partnership for sustainable development

Resource 31.10

1 With reference to Resource 31.10 and the geographic concept above, suggest reasons why Goal 17 (Global Partnerships) has been placed at the centre of the the diagram.

_____

_____

_____

_____

_____

_____

_____

_____

_____

_____

_____

_____

2 Examine the suitability of the SDGs as a framework for making the world a better place by 2030.

_____

_____

_____

_____

_____

_____

_____

_____

_____

_____

_____

_____

_____

_____

_____

_____

_____

_____

_____

_____

_____

_____

_____

_____

_____

_____

_____

_____

_____

_____

_____

_____

_____

Answers

Chapter 1

1 a i   Scattered or dispersed
     ii Linear
     iii Clustered or concentrated
   b Los Angeles has some large blended
      neighbourhoods and several smaller racially
      homogeneous neighbourhoods.
2 a i   There has been a decrease in residents aged
             20–34 in the northern coastal suburbs (e.g.
             Takapuna, Devonport, Kohimarama). There has
             been an increase in residents aged 20–34 in the
             CBD.
     ii There has been a notable increase in residents
         aged 50–64 across Auckland but most notably in
         the CBD, Whangaparaoa and the coastal suburbs.
   b Almost all areas across Auckland experienced an
      increase in residents aged 50–64 between the two
      census dates. This correlates with New Zealand's
      overall ageing population and the current age of the
      Baby Boom generation born in the 1950s and 1960s.
3 a i   Southeast of Cajuru, north of Bairro Novo
     ii Matriz
     iii Cajuru, Southwest of Portao
     iv Cambio Verde are located in close proximity
         to slums.
   b Cambio Verde as generally located adjacent to and
      within walking distance of the main slums.
4 a Teacher to mark this question.
   b Teacher to mark this question.
5 a Teacher to mark this question.
   b Teacher to mark this question.

Chapter 2

1 a i   Median rainfall
     ii Median temperature
   b i   Likelihood or probability of actual rainfall
           exceeding medium rainfall.
     ii Likelihood or probability of actual temperature
         exceeding medium temperatures.
   c The northern coastal areas of Australia have a
      less than 40 percent chance of exceeding median
      temperature during the stated period while both the
      northern and eastern coastal areas have a more than
      80 percent chance of exceeding median
      temperatures for the same period.
2 a i   90–97 girls per 100 boys
     ii 80–90 girls per 100 boys
     iii 97 or more girls per 100 boys
   b The greatest inequality in education between boys
      and girls exists in Africa and South Asia. Gender
      equality is greater in Europe and North America.
3 a i   North-eastern states, east coast, west coast.
     ii Central mid-western states.
     iii Mountain and desert regions e.g. Mohave
         Desert, Rockies.
   b Teacher to mark this question.
4 a Teacher to mark this question.
   b Teacher to mark this question.
5 a Teacher to mark this question.
   b Teacher to mark this question.
   c Teacher to mark this question.

Chapter 3

1 a Teacher to mark this question.
   b i   Estonia has stronger trade relations with Sweden
           and Finland than it does with Latvia and Lithuania.
     ii Norway has strong bilateral trade with Sweden,
         Denmark and Germany, while Sweden and
         Denmark have more diverse trade patterns.
     iii Russia is an important export country for Iceland
         and Greenland. Finland relies more heavily on

Russian trade than Sweden, Denmark or
Norway do.
2 a Teacher to mark this question.
   b The flow-line map shows that the flow of internal
      migrants away from the Auckland and Canterbury
      regions is greater than the inward flows.
   c The extent to which New Zealand's population moved
      north or south between the 2008 and 2013 is
      a matter of interpretation. Evidence to support the
      movement of people north for example could
      reference the net outflow of people from the
      Canterbury Region (particularly to Auckland). While
      arguments for the movement north could refer to the
      net outflow of Auckland's south to the Waikato.
3 a i $US 380 million
     ii $US 250 million
     iii $US 50 million
     iv $US 10 million
     v   < $US 10 million
   b Teacher to mark this question.
4 a i   Approximately 6 million
     ii Approximately 5 million
     iii Less than 100 000
     iv Approximately 1 million
   b The vast majority of slaves transported from Africa
      between 1500-1900 were sent to North and South
      America (including the Caribbean). Fewer were
      transported to North Africa, Asia and the Middle East.
   c The Transatlantic Slave Trade could have been
      considered a global issue between 1500-1900 as it
      involved the forced transportation of people between
      several continents.
5 a Teacher to mark this question.
   b Teacher to mark this question.

Chapter 4

1 a i   90 000 km³
     ii 0.2 km³
     iii 30 109 800 km³
   b i   1 200 000 km³
     ii 5 500 500 km³
     iii 1 600 000 km³
   c Highest: Antarctica. Lowest: Australia
2 a i   Approximately $130–150 billion
     ii Approximately $100 billion
     iii Approximately $300 billion
     iv Approximately $20–30 billion
     v   Approximately $500 billion
   b China has a trading relationship with almost every
      other nation in the worlds.
3 a Teacher to mark this question.
   b Teacher to mark this question.

Chapter 5

1 a Valid answers: Tropical cyclone, hurricane, cyclone,
         deep depression or low pressure system.
   b Northern Hemisphere
   c Anti-clockwise
2 a i   High pressure/anticyclone
     ii Low pressure/depression
     iii Warm front
     iv Occluded front
   b i   981 hPa
     ii 984 hPa
     iii 1028 hPa
     iv 47 hPa
   c i   Very calm, westerly to southerly breeze.
     ii Very calm, northwest to southwest breeze.
3 Teacher to mark this question.
4 The extreme contrast in weather conditions over
    a four-day period was very unusual.

Chapter 6

1 Relative location means to locate a place relative to other landmarks, while absolute location is locating a place using a coordinate system.

2

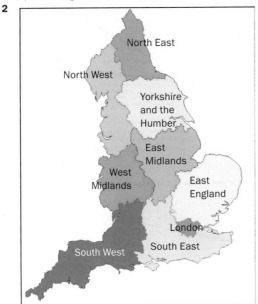

3 a 075°
   b 082°
   c 199°
   d 271°
   e 306°
4 a i SE
   ii NNE
   iii W
   iv SE
   b i 315°
   ii 95°
   iii 230°
   iv 315°

Chapter 7

1 a i GR 637075
   ii GR 613085
   iii GR 573049
   iv GR 605071 or 626091
   v GR 601095
   vi GR 586089
2 a i Golf course
   ii Skyline chalet
   iii Highway 6 bridge
   iv Bowen Peak
   v Old dam
   vi Hidden Island

Chapter 8

1 a i 1855500 E 5875700 N
   ii 1854400 E 5874900 N
   iii 1856100 E 5878700 N
   iv 1857400 E 5876600 N
   v 1853300 E 5876100 N
   vi 1852900 E 5878700 N
   b i Quarry
   ii Te Teko Rocks
   iii Airstrip
   iv Whangamata Beach
   v Hauturu Island
   vi Landfill

Chapter 9

1

Belém, Brazil	1°26' S 48°29' W
Lima, Peru	12°00' S 77°02' W
Tehran, Iran	35°45' N 51°45' E
Cairo, Egypt	30°02' N 14°08' E
Beijing, China	39°54' N 24°26' E
Tripoli, Libya	32°57' N 13°12' E
Wellington, New Zealand	41°17' S 174°47' E
Suva, Fiji	18°09' S 178°27' E
Cape Town, South Africa	33°55' S 18°25' E
Reykjavik, Iceland	64°04' N 21°58' W

2 a 35°73'21" S 174°47'25" E
   b 37°47'13" S 175°16'45" E
   c 39°29'34" S 176°54'43" E
   d 39°38'22" S 176°50'21" E
   e 40°21'08" S 175°36'30" E
   f 41°17'11" S 174°46'35" E
   g 41°16'16" S 173°17'02" E
   h 44°23'49" S 171°15'17" E
3 a Queenstown
   b Invercargill
   c Wanganui
   d Rotorua
   e Wellington

Chapter 10

1 a One unit on the map is equal to two million units on the ground.
   b One unit on the map is equal to 250 000 units on the ground.
   c One unit on the map is equal to 50 000 units on the ground.
   d One unit on the map is equal to 25 units on the ground.
   e One unit on the map is equal to 2500 units on the ground.
2 a $\frac{1}{500\,000}$ 1:500 000
   b $\frac{1}{20\,000}$ 1:20 000
   c $\frac{1}{1\,000\,000}$ 1:1 000 000
   d $\frac{1}{20\,000\,000}$ 1:20 000 000
3 a <5 hours
   b 10 hours
   c 9.5 hours
   d 10–12 hours
   e 11 hours
   f 11–15 hours
   g 21–22 hours
   h 12 hours
4 The time-scale map would give authorities an indication of how much time they had to evacuate coastal areas under threat.
5 a 5 km
   b 920 m
   c 3.48 km
   d 840 m
   e 3.33 km
   f 1.04 km
   g 360 m

Chapter 11

1
  a 147 km²
  b 3 km²
  c 32 km²
  d 4 km²
  e 9 km²

Chapter 12

1 Teacher to mark this question.
2 Teacher to mark this question.
3 Teacher to mark this question.

Chapter 13

1 Teacher to mark this question.
2 Teacher to mark this question.
3 Teacher to mark this question.
4 Teacher to mark this question.

Chapter 14

1
  a 1 730 km²
  b 293 km²
2
  a i   North America
     ii Europe and Central Asia
  b i   Europe and Central Asia
     ii Latin America and Caribbean
  c i   −138 000
     ii −1 985 000
  d 3 626 000
  e Teacher to mark this question.
  f Teacher to mark this question.

3

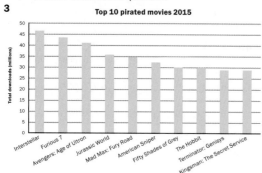

Top 10 pirated movies 2015

4

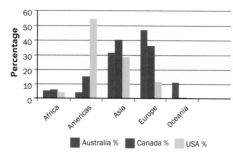

Percentage of foreign born in Australia, Canada and United States by region of birth

Chapter 15

1
  a i   25 births per 1000
     ii 13.5 births per 1000
     iii 11 births per 1000
2
  a i $11.65
     ii $12.43
     iii $93.41
  b i $0.49
     ii $32
     iii $102
  c As all earlier prices had been adjusted for inflation to a 2016 dollar equivalent.

3

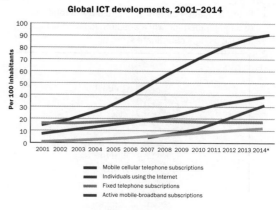

Global ICT developments, 2001–2014

4

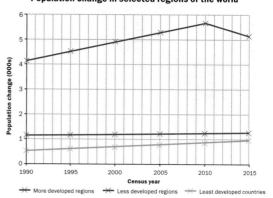

Population change in selected regions of the world

Chapter 16

1

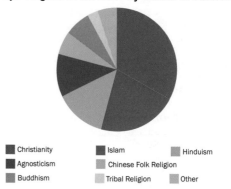

Major religions of the world by number of adherents

2 a Destination of India's electronic exports

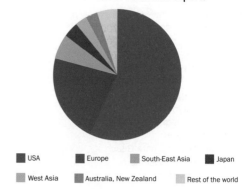

3

Greenhouse emissions by source (2016)

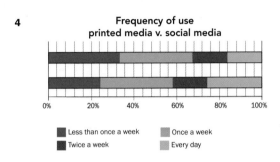

■ Power generation ■ Land use ■ Agriculture ■ Transport
■ Industry ■ Buildings ■ Waste ■ Other

4

Frequency of use
printed media v. social media

■ Less than once a week ■ Once a week
■ Twice a week ■ Every day

Chapter 17

1 a i Reading score: 539
GDP: $26 574
  ii Reading score: 449
GDP: $9000
  iii Reading score: 505
GDP: $53 672

  b There is a positive correlation between the two variables.

2

Life expectancy v. crude birth rate (2015)

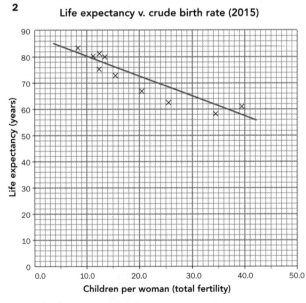

a As above.
b As the crude birth rate increases, the life expectancy of the population decreases i.e. the correlation between the two variables is negative.

3

Internet users v. % Obese

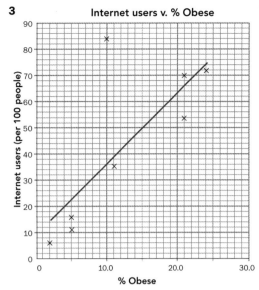

a As above.
b As the Internet use increases, obesity rates also increase, i.e. the correlation between the two variables is positive.
c Teacher to mark this question.

Chapter 18

1 a i Clay
  ii Clay loam
  iii Loam

2

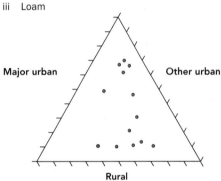

a In most Australian states, the large majority of people live in urban areas of more than 100 000 people. Exceptions to this include Tasmania and the Northern Territory where the population mainly lives in urban areas of less than 100 000.

3 a i 8.3%
  ii 12.3%
  iii 23.9%
  iv 35%
  v 42.4%
  vi 29%
  vii 11.8%
  viii 33.6%
  ix 66.3%

  b

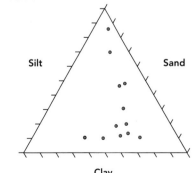

c Teacher to mark this question.

Chapter 19

1 **a** **i** 77 million

    ii 56 million

    iii 44 million

    iv 25 million

2

Vehicles per km of road

3

Military personnel

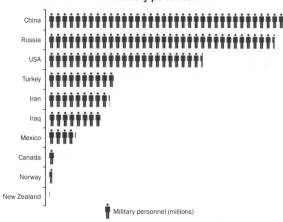

b Teacher to mark this question.

Chapter 20

1 **a** **i** Temp: 26.1°C   Rainfall: 260 mm

    ii Temp: 26.3°C   Rainfall: 300 mm

    iii Temp: 26.5°C   Rainfall: 88 mm

    iv Temp: 27.6°C   Rainfall: 126 mm

  b Wettest: March

    Driest: August

  c Warmest: October

    Coolest: February

  d Manaus experiences a hot-wet season from November through to April. During this period, both rainfall and temperature are higher than the rest of the year. Conversely, Manaus experiences a cool-dry season from June to October when both rainfall and temperature are considerably lower.

2

Climate graph for San Pedro de Atacama, Chile

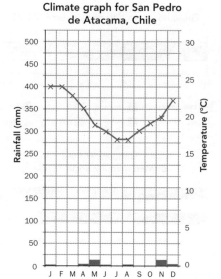

3 Teacher to mark this question.

Chapter 21

1 **a** 7%

  b 23%

  c Contracting

2

Russia (2012)

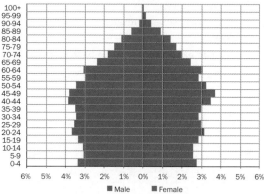

3 It has a broad top and a narrower base.

4

Brazil (2012)

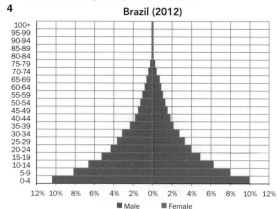

5 It has a wide base and is tapered towards the top.

Chapter 22

1 **a** **i** Resource 22.4

    ii Resource 22.2

    iii Resource 22.3

2 Both are taken vertically. However, satellite images, which are taken from space usually use artificial (or false) colours.

3 **a** The area utilised for salt mining south of the Lisan Peninsula has increased. The denser vegetation (red) to the south of the Dead Sea has all but disappeared.

b Cultural processes include salt mining and urbanisation in response to population growth.

Chapter 23

1 To simplify, identify and record important geographic features.
2 Teacher to mark this question.
3 Teacher to mark this question.

Chapter 24

1 a Teacher to mark this question.
  b Teacher to mark this question.
  c Teacher to mark this question.
  d Teacher to mark this question.
2 a Teacher to mark this question.
  b Teacher to mark this question.
  c Teacher to mark this question.
  d Teacher to mark this question.
3 a Teacher to mark this question.
  b Teacher to mark this question.

Chapter 25

1 Based on a pattern of concentric zones, Burgess' model identifies the CBD as an urban settlement's central place. Other land uses develop away from the urban centre as the settlement grows. The model is useful to geographers as it can help them to account for the form of pre-1940s urban settlements.
2 Harris and Ullman's model uses irregular-shaped nodes of activity instead of radial sectors as used by Hoyt.
3 Teacher to mark this question.

Chapter 26

1

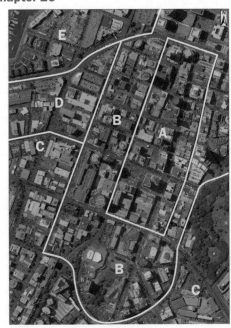

2 Teacher to mark this question.
3 Teacher to mark this question.

Chapter 27

1 Stage 1 = B
  Stage 2 = E
  Stage 3 = D
  Stage 4 = C
  Stage 5 = A
2 Teacher to mark this question.
3 Teacher to mark this question.
4 Teacher to mark this question.

Chapter 28

1 a i Core: UK, France, Germany, Italy, Denmark, Sweden

ii Periphery: Portugal, Croatia, Poland, Latvia, Lithuania, Greece
iii Semi-perphery: Spain, Finland, Czech Republic
  b Teacher to mark this question.
2 Better employment prospects, higher standard of living.

Chapter 29

1 a Auckland 15.2°C   Dunedin 11.1°C
  b Auckland 14°, 16°, 19°   Dunedin 7°, 9°, 12°, 14°, 15°
  c Auckland 10°   Dunedin 8°
2 a 57.4 years
  b 52 years
  c 15 years

Chapter 30

1 a 95.1%   b 31.0%
  c −12.8%   d 40.5%
  e 14.9%   f −20.2%
  g 136.5%   h 175.0%

Putting it all together

Development Goal: Zero hunger

1 a Teacher to mark this question.
  b Higher rates of undernourishment affect sub-Saharan nations (e.g. Central African Republic = 48%). The nations of North Africa have lower rate of undernourishment (e.g. Morocco and Algeria = 5%).
  c Teacher to mark this question.
2 a

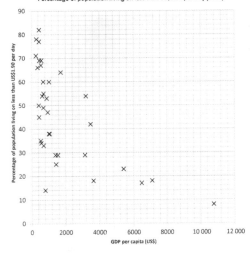

  b Increased earnings (GDP per capita) leads to lower rates of poverty.
3 Response must draw upon evidence presented in the previous two questions. Named examples should be quantified.

Development Goal: Quality education

1 a

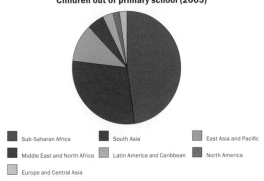

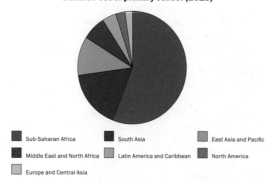

Children out of primary school (2013)

Sub-Saharan Africa South Asia East Asia and Pacific

Middle East and North Africa Latin America and Caribbean North America

Europe and Central Asia

b   The proportion of primary-aged children out of school has increase in Sub-Saharan Africa relative to the rest of the world. South Asia has experienced the greatest decrease in the 10-year period (29% to 17%) relative to the rest of the world.

c   South Asia has experienced the greatest progress while Sub-Saharan Africa has seen the least relative to the rest of the world.

Development Goal: Gender equality

1 a   The proportion of women represented in the European Parliament has steadily increased in the period between 1979 and the current term.

   b   Teacher to mark this question.

Development Goal: Good health and well-being

1 a

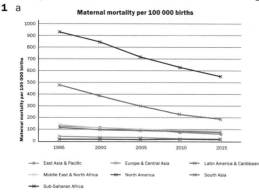

Maternal mortality per 100 000 births

East Asia & Pacific   Europe & Central Asia   Latin America & Caribbean

Middle East & North Africa   North America   South Asia

Sub-Saharan Africa

b   Maternal mortality has decreased in all regions. Responses should refer to named regions and supported by quantification of data.

2 a   Sub-Saharan Africa

   b   North America, Europe and Australasia

   c   Teacher to mark this question.

Development Goal: Sustainable cities and communities

1   Teacher to mark this question.

Development Goal: Industry, Innovation and Infrastructure

1 a   The global core countries of Europe and North America are well-connected. Southern and Eastern Asia are also well-connected. South America is moderately connected while connections in Africa, the global periphery, are least connected.

   b   Wealthier regions of the global core are more connected than the poorer regions of the global periphery.

Development Goal: Responsible consumption and production

1 a   In the first frame, the consumer is affected by the high oil price of $100. In the second frame, the 'green revolution' is affected by lower oil price of $60 or less.

   b   Refers to the drop in oil prices, which affects consumer behaviour and may hinder growth of green technologies.

Applying geographic concepts: Interaction

1   Teacher to mark this question. Responses to this question should focus on the geographic concept of interaction and identify that a change in one element of the diagram will likely lead to a change in another.

2   Teacher to mark this question. Responses should focus on the suitability of the SDGs specifically covered in the previous questions as a framework for making the world a better place. Answers to previous questions should be used as supporting evidence in one's argument.

ISBN: 9780170389341